PAINLESS

Math Word Problems

Marcie F. Abramson
with contributions by Rika Spungin
illustrated by Laurie Hamilton

BARRON'S

To my editor, Linda Turner, for her abundant guidance.

To my friend and colleague, Rika Spungin, for saving the day.

To my brothers, Howard and Jerry, for their amazing help with my new friend, the computer.

And, of course, to my parents, Cynthia and Charles Abramson, for their continued love, support, and encouragement in all of my endeavors.

All inquiries should be addressed to:
Barron's Educational Series, Inc.
250 Wireless Boulevard
Hauppauge, New York 11788
http://www.barronseduc.com

Library of Congress Catalog Card No. 00-031248
International Standard Book No. 0-7641-1533-2

Library of Congress Cataloging-in-Publication Data

Abramson, Marcie F.
    Painless math word problems / Marcie F. Abramson ; illustrated by Laurie Hamilton.
        p.   cm.
    Includes index.
    ISBN 0-7641-1533-2
    1. Problem solving—Juvenile literature. 2. Word problems
        (Mathematics)—Juvenile literature. [1. Word problems
        (Mathematics)] I. Title: Word problems. II. Hamilton, Laurie,
        ill. III. Title.
    QA63 . A32   2001
    510'.76—dc21   00-031248

PRINTED IN THE UNITED STATES OF AMERICA
9 8 7 6 5 4 3 2 1

CONTENTS

INTRODUCTION

Painless word problems? How can that be? Word problems
aren't scary! Just wait and see!

After many years of teaching, I have discovered that when the
expression "word problems" is mentioned, groans and moans
can be heard from near and far. Since life is full of word problems that need to be solved, I decided to write a book on how
painless word problems can be when they are solved carefully.
With some easy tips, lots of examples and practice, and a bit of
humor, you will be able to solve all types of word problems, and,
hopefully, never groan again at the mention of word problems!

Chapter One will take the "problem" out of "word problems."
Learn how to read and understand any problem and then make a
plan for solving it.

Chapter Two leaps into word problems that contain whole
numbers. Add, subtract, multiply, and divide your way to easy
solutions to big problems. Learn how to see patterns, make lists,
and guess and check your way to an answer, too.

Chapter Three dives into decimals and fractions. Learn how
to compute with decimals and fractions as you solve real-life
word problems.

Chapter Four runs into rates and proportions. Learn how
comparing rates and solving proportions can be done with ease.

Chapter Five plows into percents. Learn how painless it can
be to solve sales, banking, price increase, and other problems by
using percent tips.

Chapter Six soars into statistics and probability problems.
Learn how to use mean, median, range, and mode in everyday
real-world problems. See how probability is used to predict
future outcomes and events.

Chapter Seven jumps into geometry and measurement. Learn
how to find perimeter, circumference, and area in real-world
problems. Become a whiz at solving all types of measurement
problems. Learn about triangles and using the Pythagorean
theorem.

Chapter Eight enters into equations and a bit of algebra.
Learn how equations can be used to solve everyday problems in
a quick and easy way.

Chapter Nine is full of practice problems. Use your new skills and you will see, that solving word problems can be done painlessly!

Chapter Ten engages the world of the Internet. Surf, search, and solve word problems and activities that can be found on the World Wide Web.

Web Addresses Change!

You should be aware that addresses on the World Wide Web are constantly changing. Although the addresses provided were current when this book was written, sooner or later, some of the addresses may no longer work. If you should come across a web address (URL) that no longer appears to be valid, either because the site no longer exists or because the address has changed, either **shorten the URL to the first slash** or do a **key word search** on the subject matter or topic.

To do a key word search, if you're interested in finding out how math is used in everyday life, for example, try typing in "Daily Math." Don't forget the quotes! From there, you can go to cooking or home decorating or even banking. These are all key words. Remember to use key words that will narrow your search to the specific topic you are looking for.

WARNING: Not every response to your search will match your criteria, and some sites may contain adult material. If you are ever in doubt, check with someone who can help you.

CHAPTER ONE

Taking the Problem Out of Word Problems

A word problem is no problem when four steps you do:

Read it,
Plan it,
Solve it,
Check it.

These four steps make a problem painless for you!

FOUR STEPS TO WORD PROBLEM SUCCESS

The biggest problem with a word problem is deciding how to solve it.

In his book *How to Solve It*, George Polya (1887–1985), a famous mathematician, described four steps that can make solving word problems easier. Here are the four steps in easy-to-understand language:

Step 1: Understand the problem.

Step 2: Plan a strategy.

Step 3: Do the plan.

Step 4: Check your work.

Understanding the problem means reading the problem carefully so that you know what is being asked in the problem. What is the question? Is there enough information in the problem for it to be answered?

Planning a strategy means deciding what to do to solve the problem. You may need to add, subtract, multiply, or divide to find the answer. Perhaps you will need to draw a picture of what is given, or make a list of possible answers in order to find the right answer, or even guess the answer and check it to see if it is right. There are many strategies you will learn that will be explained and used as we go on.

Doing the plan means carrying out your strategy.

Checking your work means rereading and redoing the problem to make sure that your work is mathematically accurate and that your answer makes sense.

STEP 1: UNDERSTAND THE PROBLEM

To understand the problem you will need to read it carefully and then ask yourself these three questions:

1. What is the problem about?

2. What are you being asked to find?

3. Is there enough information given to solve the problem?

When you read a problem and can answer all three of these questions you will be ready to start solving word problems with ease.

REMEMBER!

It is important to read *all* of the words in a word problem. Try not to concentrate just on the numbers mentioned or the words that appear at the end of the problem. You may miss some important information and may not be able to solve the problem.

EXAMPLE:

Stella Stockman bought 200 shares of Kooky Kola stock at $15 per share. Not including any fees, what was the cost of the stock?

Step 1: Understand the problem.

What is the problem about? buying shares of stock

What are you being asked to find? the total cost of Stella's stock

Is there enough information given to answer the question? Yes; you know how many shares she wants to buy and the cost of each share of stock. (NOTE: You will learn about deciding when to add, when to subtract, when to multiply, and when to divide later in this chapter.)

MATH NOTE

Sometimes a word problem is not stated in question form, but rather with a statement that represents the question being asked.

EXAMPLE:

Charlie Chipper baked 48 cookies for himself and for his friends. Find how many cookies each person will receive if the cookies are shared equally.

Step 1: Understand the problem.

What is the problem about? baking and sharing cookies with friends

What are you being asked to find? how many cookies each person will receive

Is there enough information given to answer the question? No; you do not know the number of friends.

BRAIN TICKLERS
Set # 1

To understand each word problem, answer the following questions:

a. What is the problem about?
b. What are you being asked to find?
c. Is there enough information given to answer the question?

It is not necessary to solve the problem. You will solve many problems later after you learn all four steps to word problem success!

1. Felicia Fielder has 75 more baseball cards in her collection than Ben Basehart has in his collection. Ben has 248 cards in his collection. How many cards are in Felicia's collection?

2. Manual Musicler bought 15 CDs at a total cost of $195 (not including the tax.) Find the cost of each CD if each CD cost the same amount.

3. Portia, Pedro, and Paige went on a vacation. Portia snapped a total of 17 rolls of film, Paige snapped a total of 20 rolls of film, and Pedro snapped a total of 15 rolls of film. If there were 24 pictures per roll, how many pictures did Portia snap?

Caution—Major Mistake Territory!

Sometimes, extra information is given that is not needed to solve the question being asked. In this case, you do not need to know how many rolls Paige and Pedro snapped in order to answer the question.

4. There are 29 players on the Blue Sox softball team, 25 players on the Halibuts softball team, and 30 players on the Planes softball team. Find the total number of players on the three teams.

5. The Peppy Pizza Parlor baked 15 pizzas for the New Town School picnic. Each pizza was cut into eight slices. Each person at the picnic ate one slice of pizza, and none was left over. How many people were at the picnic?

MATH NOTE

Read carefully! Sometimes, numbers are spelled out as words, such as "eight" for "8."

6. Curtis Carter parked his car at the Corner Lot. The lot charges $5 an hour or part of an hour. Curtis parked his car for $4\frac{1}{2}$ hours. What was Curtis' total cost for parking his car at the Corner Lot?

7. Anica Archer is learning the sport of archery. In her third lesson, she shot five arrows, three of which hit the target. One arrow hit the gold ring, one arrow hit the white ring, and one arrow hit the black ring. How many points did Anica score in her third lesson?

8. Victor's Video Arcade pays its employees $7.95 per hour for weekday hours and $9.25 for weekend hours. Chan worked seven hours on Sunday. How much money did Chan earn working on Sunday?

9. The Science Museum charges a $6.25 admission fee for each person aged 12 years or older. Children under 12 are admitted free of charge. If $7,500 in admission tickets were sold on Sunday, how many people aged 12 or older visited the museum that day?

10. The lightbulb was invented in 1879, and the movie camera was invented in 1889. The electric vote recorder was invented eleven years before the lightbulb was invented. In what year was the electric vote recorder invented?

(Answers are on page 18.)

STEP 2: PLAN A STRATEGY BY CHOOSING THE OPERATION

A word problem is no problem when you have a plan:
Choose a strategy,
Work it out;
You did it! You deserve a hand!

For any word problem, read the problem carefully to decide which operation makes sense for the problem situation.

Using addition

You use addition to

- tell how many altogether;
- tell how many in all;
- find the total number (amount, cost).

Using subtraction

You use subtraction to

- tell how many are left;
- tell how much is left;
- tell how many more;
- tell how many fewer;
- tell how much more;
- tell how much less;
- tell how many more are needed;
- tell how many are not . . . .

Using multiplication

You use multiplication when you know the number of groups and the number (amount) in each group to

- tell the total number (amount);
- to tell how many altogether;
- to tell how many in all.

Using division

You use division when

- you know the total number (amount) and the number of groups to tell the number (amount) in each group;
- you know the total number (amount) and the number (amount) in each group to tell the number of groups.

Perhaps you can think of other situations that imply addition, subtraction, multiplication, or division. That's great! There are so many math phrases for each of the operations that we can only list some of them here. Keep track of words and other phrases that you discover that imply different operations. This will help you solve many types of word problems.

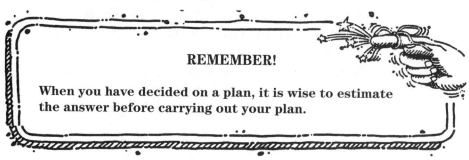

REMEMBER!

When you have decided on a plan, it is wise to estimate the answer before carrying out your plan.

EXAMPLE:

President Harry Truman made the first televised presidential address. His telecast occurred on October 5, 1947. The first televised presidential debate occurred 13 years after Truman's address. In what year was the first televised debate?

Step 1: Understand the problem.

What is the problem about? presidential addresses and debates

What are you being asked to find? the year of the first debate

Is there enough information given to answer the question? Yes; you know the year of the address and the number of years later that the first televised debate occurred.

Step 2: Plan a strategy.

What operation is suggested? addition, since the year of the debate is 13 years *after* Truman's televised debate

Estimate the year. later than 1947

9

EXAMPLE:

Little Golden Books were first introduced in 1942 at a cost of $0.25 for one book. Find the total cost of buying 17 of the original *Little Golden Books* in 1942 (tax not included).

Step 1: Understand the problem.

What is the problem about? *Little Golden Books*
What are you being asked to find? the cost of 17 of the original books
Is there enough information given to answer the question? Yes; you know the number of books bought and the cost of each book.

Step 2: Plan a strategy.

What operation is suggested? multiplication, since you can use the number of books and the cost of one book to find the total cost
Estimate the cost of the 17 books. more than $4.00:

$17 \times 0.25 = 17 \times \dfrac{1}{4}$, which is more than $16 \times \dfrac{1}{4} = \$4.00$.

MATH NOTE

Did you read the problem above and decide to use addition? You could add $0.25 seventeen times, but multiplying $0.25 by 17 is easier. Multiplication is really repeated addition, but it is usually easier to do one multiplication step than to do many addition steps.

EXAMPLE:

A $1 bill measures 6.14 inches long. How many $1 bills placed end to end would stretch the length of a football field 100 yards long? (Remember that one yard = 36 inches.)

Step 1: Understand the problem.

What is the problem about? dollar bills in a line 100 yards long
What are you being asked to find? how many bills placed end to end would make a length of 100 yards?

Is there enough information given to answer the question? Yes; you know the length of one bill and the length of the football field.

Step 2: Plan a strategy.

What operation is suggested? division, since you know the total length and the length of one bill

Estimate the answer. about 600 (100 yards = 3600 inches; the bills are about 6 inches long, and 3600 inches ÷ 6 inches for each bill = 600 bills.)

MATH NOTE

Sometimes drawing a picture will help you solve a problem. This may help you to decide if you need to add, subtract, multiply, or divide. For the previous Example, you could draw the following picture:

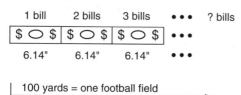

By looking at the picture, you can see that you are dividing a length of 100 yards into single sections of 6.14 inches each. So, you should divide! To solve the problem, you need to change the 100 yards into its corresponding equivalent length in inches using the fact that there are 36 inches in one yard.

BRAIN TICKLERS
Set # 2

For each word problem, answer the following:

a.   What is the problem about?
b.   What are you being asked to find?
c.   Is there enough information given to answer the question?
d.   What operation is suggested?
e.   Estimate the answer.

If you cannot decide on the operations, you may need to draw a picture to help you.
It is not necessary to solve the problem.

Note that Problems 7 to 9 are repeated from the first set of Brain Ticklers, which begins on page 5. For these problems you need only do parts d and e above.

1. Volleyball became an official Olympic sport in 1964. The sport of basketball was invented 73 years before volleyball became an Olympic sport. In what year was basketball invented?

2. Big Ben, located in London, England, is the name of the largest bell in its clock tower. The bell weighs 37.5 tons. There are 2,000 pounds in a ton. How many pounds does the Big Ben bell weigh?

3. How many 0.5-centimeter long beads are needed to create a bracelet that will be 20 centimeters long?

4. The first medical school for women opened in Boston, Massachusetts, in 1848. The United States granted women the right to vote 72 years after the Boston medical school opened. In what year did women receive the right to vote?

5. The phone number for the White House in Washington, D.C., is 201-456-1414. This is not the President's personal line but a phone number from which information about tours, hours of operation, and other general information may be obtained. A 5-minute weekday afternoon call from New York City to the White House under one calling plan costs $0.75. Find the cost per minute of phoning the White House from New York City on a weekday afternoon.

6. There are 11 Life Saver candies in each standard-size roll of Life Savers. How many Life Saver candies would you have in all if you bought 12 rolls?

7. The Science Museum charges a $6.25 admission fee for each person aged 12 years or older. Children under 12 are admitted free of charge. If $7,500 in admission tickets were sold on Sunday, how many people aged 12 or older visited the museum that day?

8. Felicia Fielder has 75 more baseball cards in her collection than Ben Basehart has in his collection. Ben has 248 cards in his collection. How many cards are in Felicia's collection?

9. The lightbulb was invented in 1879, and the movie camera was invented in 1889. The electric vote recorder was invented eleven years before the lightbulb was invented. In what year was the electric vote recorder invented?

(Answers are on page 19.)

MULTI-STEP PROBLEMS

A word problem is no problem
Whether it needs one step or maybe two.
Choose the operations, then estimate the answer;
This will make it painless for you!

Some word problems require two (or more) steps in order to find the answer to the problem. If this is true, choosing and sequencing the operations, estimating the answer, and checking your answer are very important steps to follow.

EXAMPLE:
Jeffrey bought four bags of potato chips.
Each bag cost $1.49. He also bought
three bottles of soda. Each bottle
cost $1.29. Find the total cost
of his purchase.

Step 1: Understand the problem.
What is the problem about? buying chips and soda
What are you being asked to find? the total cost of the
purchase
Is there enough information given to answer the question?
Yes; you know the number of each item purchased and
the cost of each item.

Step 2: Plan a strategy.
What operations are suggested? multiplication to find
the total cost of the potato chips; a second multiplication to find the total cost of the bottles of soda; then addition to find the total cost of both the chips and the
soda.
Estimate the answer. more than $8.00 (The potato chips
cost more than $5.00, and the bottles of soda cost more
than $3.00.)

EXAMPLE:
Skyway Travel Company offers a weekend package to Disneyland that includes airfare and hotel. The cost is $450 for each
adult and $189 for each child aged 17 or under. A family of two
adults and three children, each under age 17, purchased the
package. What was the total price of the family's weekend package to Disneyland?

Step 1: Understand the problem.
What is the problem about? a package trip to
Disneyland
What are you being asked to find? the total price of the
family's weekend package to Disneyland

Is there enough information given to answer the question? Yes; you know the cost of the adult and children's tickets, and how many adults and children went on the trip.

Step 2: Plan a strategy.

What operations are suggested? multiplication to find the total cost of the adult tickets; a second multiplication to find the total cost of the children's tickets; then addition to find the total cost of all tickets.

Estimate the answer. about $1,400 (The adult tickets cost a total of more than $800, or 2×400; the children's tickets cost a total of less than $600, or 3×200; and $800 + $600 = $1,400.)

EXAMPLE:

Your height is about six times the length of your foot. Samara's foot measures ten inches in length. About how many feet tall is Samara? (Remember that 12 inches = 1 foot.)

Step 1: Understand the problem.

What is the problem about? Samara's foot length and her height

What are you being asked to find? Samara's approximate height in feet

Is there enough information given to answer the question? Yes; you know the number of times greater Samara's height is than her foot length, and you know the length of her foot in inches. You also know the number of inches in a foot.

Step 2: Plan a strategy.

What operations are suggested? multiplication to find Samara's height in inches; then division, since you know her total height in inches and the number of inches in one foot.

Estimate the answer. A reasonable (common sense) estimate is between 4 feet and 6 feet.

MATH NOTE

Once again, drawing a picture of a problem can help you decide what operations to use. For the previous example, you can do the following:

1. Draw a length to represent Samara's foot length. Repeat this length six times to represent Samara's height.

| 10 inches | 10 inches | 10 inches | 10 inches | 10 inches | 10 inches |

You can add the lengths that you have drawn or multiply each foot's length by how many lengths you have drawn.

2. To find the number of feet, divide the total number of inches by 12. (Finding the number of 12-inch pieces given a total length in inches implies division).

BRAIN TICKLERS
Set # 3

In this set of Brain Ticklers, enough information is given to solve each multi-step word problem.

For each problem, answer the following:

a.   What is the problem about?
b.   What are you being asked to find?
c.   What operations are suggested?
d.   Estimate the answer.

After you have answered a to d above, list, in your own words, the steps you would use to solve the problem. It is not necessary to find the answer.

1. Movie Time Cinema charges $8.50 per person aged 13 or older and $4.25 per child under age 13. Friday night, Movie Time took in a total of $850 in children's tickets and $1,275 in tickets to persons 13 or older. How many more tickets were sold to children under age 13 than were sold to persons 13 or older?

2. ABC stock opened at $12\frac{1}{8}$ points on Monday morning. By 10:00 AM, it had gained $1\frac{1}{4}$ points. Between 10 AM and 1:00 PM the stock lost $\frac{1}{16}$ of a point. Between 1:00 PM and 3:00 PM, the stock increased in price by $\frac{1}{2}$ of a point, but then decreased by $\frac{3}{4}$ of a point between 3:00 PM and the close of the trading day. What was the closing price of ABC stock on Monday?

3. Graham Gardner planted gardenias in one line of his garden. He planted one gardenia every 18 inches for a total length of 10 feet. How many gardenias did Graham plant in the line? (HINT: Drawing the line of gardenias will help you solve the problem.)

4. The M&M/Mars Company states that in a bag of 500 plain M&M's, about 100 of the candies are red and about 50 of the candies are green. In a bag containing 500 candies, about how many candies are not red or green?

5. A Monopoly playing board is built in the shape of a square. Each side has 11 landing places. If you travel around the board five complete times, over how many landing places will you travel? (HINT: Draw a diagram of the playing board with 11 landing places on each side.)

(Answers are on page 21.)

We've read our problems carefully,
And learned how to make a plan;
By choosing the operations and estimating,
The problems we understand.
The time has come
To do the plan and see
Whether by computing or drawing a picture
An answer there will be.
In the next chapters we will try
To continue our mathematical quest
To read, to plan, to do, to check
Our way to word problem success!

BRAIN TICKLERS—THE ANSWERS

Set # 1, page 5

NOTE: Some of your answers may be written differently from the answers stated below. This is all right! Just make sure that your answers and the book's answers have the same meaning.

1. a. baseball card collections
    b. the number of cards in Felicia's collection
    c. Yes; you know the number of cards in Ben's collection and how many more cards are in Felicia's collection.

2. a. buying CDs
    b. the cost of each CD
    c. Yes; you know the total cost and the number of CDs.

3. a. snapping rolls of film
    b. how many pictures Portia snapped
    c. Yes; you know the number of rolls she snapped and the number of pictures per roll.

4. a. softball teams
    b. the total number of players on all the teams
    c. Yes; you know the number of players on each softball team.

5. a. a pizza picnic
    b. how many people were at the picnic
    c. Yes; you know the total number of pizzas, how many slices each person ate, and that no slices were left over.

6. a. parking at the Corner Lot
    b. the cost to park for $4 \frac{1}{2}$ hours
    c. Yes; you know the cost per hour and the total hours parked.

7. a. the sport of archery
    b. Anica's score
    c. No; you do not know the point value of each ring.

8. a. working at a video arcade
    b. how much Chan earned on Sunday
    c. Yes; you know the pay rate for weekend hours and the number of hours Chan worked on Sunday.

9. a. visiting the Science Museum
    b. how many people aged 12 or older visited the museum on Sunday
    c. Yes; you know the total amount collected and the cost of each ticket bought by a person 12 or older.

10. a. inventions
    b. the year in which the electric vote recorder was invented
    c. Yes; you know when the lightbulb was invented and how many years before then that the electric vote recorder was invented.

Set # 2, page 12

1. a. volleyball and basketball
    b. the year that basketball was invented
    c. Yes; you know the year in which volleyball became an Olympic sport and how many years before then that basketball was invented.
    d. "Years before" implies subtraction.
    e. before 1900 (1964 − 73 < 1900)

2. a. Big Ben
   b. the weight, in pounds, of Big Ben
   c. Yes; you know the weight of Big Ben in tons and the number of pounds in one ton.
   d. multiplication, since the number of pounds in a ton times the number of tons gives the total number of pounds
   e. between 60,000 and 80,000 ($30 \times 2,000 = 60,000$; $40 \times 2,000 = 80,000$)

3. a. beads in a bracelet
   b. the number of beads needed to make a bracelet
   c. Yes; you know the length of each bead and the length of the bracelet.
   d. division, since you know the total length of the bracelet and the length of each bead (A picture will also show that you need to divide to find the number of beads needed.)
   e. more than 20 (If each bead were one centimeter long, there would be 20 beads in the bracelet. Since the beads are less than one centimeter long, more beads are needed.)

4. a. the first medical school for women and when women received the right to vote
   b. in what year women received the right to vote
   c. Yes; you know the year of the opening of the medical school and how many years later women received the right to vote.
   d. "72 years later" implies addition.
   e. later than 1850

5. a. phoning the White House from New York City
   b. the cost, per minute, of a weekday afternoon phone call
   c. Yes; you know the total cost of the call and how many minutes the call lasted.
   d. division, since you know the total cost and the number of minutes the call lasted.
   e. less than 20¢ a minute (20¢ a minute $\times 5$ minutes $= \$1.00$; the call was less than $1.00.)

6. a. Life Saver candies
   b. the number of candies in 12 rolls
   c. Yes; you know the number of candies in one roll and the number of rolls bought.
   d. multiplication, since you know the number of groups (rolls) and the number in each group (roll)
   e. more than 120 candies (If there were only 10 candies in a roll, there would be 120 candies altogether in 12 rolls.)

Note that the answers to Parts a, b, and c of Problems 7 to 9 are stated in the answers to the Brain Ticklers, Set # 1, Problems 9, 1, and 10, on pages 18 and 19.

7. d. division, since you know the total amount and the cost of one ticket
   e. more than $1,000 (If tickets were $6.00 each and 1,000 tickets were sold, only $6,000 in tickets would have been sold.)

8. d. addition, since you know how many cards Ben has and the number more that Felicia has
   e. more than 300 cards

9. d. "Eleven years before" implies subtraction.
   e. before 1870

Set # 3, page 16

1. a. Movie Time Cinema and its ticket prices
   b. how many more children's tickets were sold than adult tickets
   c. division to find the number of adult tickets sold; division to find the number of children's tickets sold; then subtraction to find how many more children's tickets were sold
   d. fewer than 100 more children's tickets (About $800 \div 4 = 200$ children's tickets were sold, and more than $1,000 \div 10 = 100$ adult tickets were sold.)

2. a. ABC stock
   b. the closing price of ABC stock on Monday
   c. addition to find each increase; subtraction to find each decrease
   d. fairly close to the original value, since the gains and losses were relatively small

3. a. Graham's garden
   b. the number of gardenias in one line
   c. multiplication to find the total number of inches in 10 feet; division to find how many groups of 18 inches are in the line (NOTE: Remember to add one to the quotient to account for the first flower. Drawing a picture will help you see why you need to divide the total length by the distance between each gardenia, and why you need to add one to account for the first flower.)
   d. fewer than 10 ($18 \times 10 = 180$; $180 > 120$)

4. a. M&M candies
   b. the number of candies in a bag that are not red or green
   c. addition to find how many are red or green; subtraction to find how many are not red or green
   d. more than 250 (Fewer than half of the candies are red or green.)

5. a. a Monopoly board
   b. over how many landing places you will travel in five circuits
   c. "Times" implies multiplication. (NOTE: It is helpful to draw a picture so that you can count the number of squares that you pass over when you travel around the board once. Otherwise, you may decide that there are 11 places times 4 sides on the board, or 44 places on the board. This is not true! In a picture you can see that each corner place lies on two sides but should only be counted once. So, there are actually $44 - 4 = 40$ places on the board.)
   d. fewer than 44 (Corner places are counted on two sides.)

CHAPTER TWO

Happy Whole Numbers

A word problem with whole numbers
Can be solved 1, 2, 3:
Do your plan, check it, too,
And compute accurately.

WHOLE NUMBER OPERATIONS

In approaching word problems, you have learned to use the first two of Polya's problem-solving steps: Step 1: Understand the problem, and Step 2: Plan a strategy. In this chapter, you will have more experience with Step 2. You will also learn to use Step 3: Do the plan, and Step 4: Check your answer.

You will notice that some of the problems and examples from Chapter One appear in this and future chapters. Also notice that we will no longer list the three questions that must be answered for Step 1: What is the problem about? What are you being asked to find? Is there enough information? But, don't forget to ask yourself these questions before planning a strategy, estimating the answer, and solving each problem.

EXAMPLE:

Volleyball became an official Olympic sport in 1964. The sport of basketball was invented 73 years before volleyball became an Olympic sport. In what year was basketball invented?

Step 2: Plan a strategy.

What operation is suggested? "Years before" implies subtraction.

Estimate the year. before 1900 ($1964 - 73 < 1900$)

Step 3: Do the plan.

Basketball was invented 73 years before volleyball became an official Olympic sport in 1964. Therefore, subtract 73 from 1964.

$$1964 - 73 = 1891$$

Step 4: Check your work.

One way to check your answer is to work backwards. Working backwards means that you must use *inverse operations.* The basic operations and their corresponding inverse operations are shown in the following table.

Operation	Inverse operation for checking
addition	subtraction
subtraction	addition
multiplication	division
division	multiplication

To solve this problem, you used subtraction. Therefore, to check your answer use addition. To check that $1964 - 73 = 1891$, add.

$$1891 + 73 = 1964$$

Since 1964 is the year volleyball became an official Olympic sport, 1891 is the correct answer. The check works. Basketball was invented in 1891.

EXAMPLE:
Samara's height is about six times the length of her foot. If her foot measures about 10 inches long, about how many feet tall is Samara?

Step 2: Plan a strategy.
What operations are suggested? multiplication to find Samara's height in inches; then division, since you know her total height in inches and the number of inches in one foot
Estimate Samara's height. A reasonable estimate is between 4 feet and 6 feet.

Step 3: Do the plan.
First, multiply Samara's foot length by 6.

$$10 \text{ in.} \times 6 = 60 \text{ in.}$$

But wait! The question asks for the height in feet and we found inches! So, change 60 inches to feet, remembering that there are 12 inches in one foot.

$$60 \text{ in.} \div 12 \text{ in. per ft} = 5 \text{ ft}$$

Step 4: Check your work.

Work backwards. First, check your second operation. Division is the operation, so check with multiplication.

5 ft × 12 in. per ft = 60 in.

This is correct. Samara is 60 inches tall.
Second, check your first operation. Multiplication is the operation, so check with division.

60 in. ÷ 6 = 10 in.

This is also correct. Samara's foot is about 10 inches long.
Your answer to the problem is correct. Samara is about 5 feet tall.

BRAIN TICKLERS
Set # 4

Solve each word problem by reading it carefully and thoroughly, planning a strategy, estimating the answer, doing the plan, and then checking your work. Then, in your own words, list the steps you used to solve the problem.

The solution box following the problems contains the answers to the problems. As you answer each problem, cross out the answer in the solution box. Use the number that is not crossed out to answer the following question:

For the movie The Wizard of Oz, how many pairs of shoes were dyed emerald green for the final procession?

1. Felicia Fielder has 75 more baseball cards in her collection than Ben Basehart has in his collection. Ben has 248 cards in his collection. How many cards are in Felicia's collection?

2. There are 24 players on the Quails softball team and 27 players on the Ducks softball team. There are 18 fewer players on the Buzzards football team than on both softball teams combined. Find the number of players on the Buzzards football team.

3. Jacob Jogger and Lisa Lap entered an 8-mile walkathon to raise money for a local food pantry. Jacob had 12 sponsors who each pledged $4 per mile, and Lisa had 18 sponsors who each pledged $2 per mile. If Jacob and Lisa both walked the entire route, how much more money did Jacob receive from his sponsors than Lisa received from her sponsors?

4. Boris Banker had $1,000 in his checking account. On Monday he made a deposit of $250, and on Thursday he made a withdrawal of $540. After his withdrawal, how much money was in Boris' account?

5. A cubit is an ancient measure found by measuring the distance from a person's elbow to the tip of the middle finger. Josh's cubit measures 20 inches long. How many of Josh's cubits would it take to form a line reaching one mile long? (1 mile = 5,280 feet)

6. Kareem Korrect scored 73, 85, 81, and 71 on his last four math tests. He has one more math test. A total score of 400 on the five tests will give Kareem an average of 80. What must Kareem score on the fifth test to have an average of 80?

7. A parking garage charges $6 for the first hour and $5 for each additional hour or part of an hour. If you park your car from 1:00 PM to 5:30 PM, what will be the parking cost?

8. Sorelle Songstreet bought a new stereo system. She gave a $25 deposit and agreed to pay the remainder in installments of $15 per month for the next two years. What is the total cost of the stereo system?

9. A class of sixth graders sold school stationery as a fundraiser. Boxes of notepaper cost $12 per box, and boxes of note cards cost $8 per box. The class sold a total of 55 boxes of notepaper and collected a total of $1,236 for all stationery sold. How many boxes of note cards did the class sell?

10. Art Auto and his three friends bought a used car for $1,250. They fixed it up by spending $575 on parts and repairs. They then sold the car for $4,605. If Art and his friends shared the profit equally, how much of the profit did each person receive?

Solution Box

33	90	
710	3,168	385
26	695	72
323	300	96

(Answers are on page 50.)

WORD PROBLEMS: FIND THE INFORMATION

A word problem's information
May be found in a display,
A chart, a table, or a list.
Search and solve for an answer today!

Have you ever searched a menu for favorite foods and then found the total cost of your meal? If so, you have already learned another word problem strategy: finding data (numerical information) needed to solve a problem in a display.

Data can be displayed in many forms. Here are some examples: calendars, menus, train schedules, television guides, sales order forms, recipes, banking forms, and highway signs. Think of other ways in which you can see data displayed. The world is full of charts and tables, and it is important to know how to read them.

EXAMPLE:

Zuzu ordered a Megaburger and french fries. George ordered a hamburger and an apple pie. Use the chart below to find how many more calories were in Zuzu's order than in George's order.

Burger Buddy Nutrition Guide

Item	Calories	Grams of fat
Megaburger	640	39
Hamburger	260	10
French fries	372	20
Chicken tenders	236	13
Apple pie	320	14

Step 2: Plan a strategy.

What operations are suggested? addition to find the total number of calories in Zuzu's order; addition to find the total number of calories in George's order; then subtraction to find how many more calories were in Zuzu's order than George's order.

Estimate the answer. about 400 calories (Zuzu's order has over 300 + 600 = 900 calories; George's order has over 200 + 300 = 500 calories; and 900 – 500 = 400.)

Step 3: Do the plan.

Look at the chart and its three columns titled *Item*, *Calories*, and *Grams of fat*. Which columns will we need to solve the problem? Since the question deals only with food items and the number of calories in each item, we need use only the first two columns.

For Zuzu's order, look down the Item column for the Megaburger. Look straight across to the Calories column to find the number of calories in the Megaburger (640). A Megaburger has 640 calories. Doing the same for french fries, you find that they have 372 calories.

To find the total calories in Zuzu's order, add: 640 + 372 = 1,012.

For George's order, look down the Item column for the hamburger. Looking straight across to the Calories column you find that a hamburger has 260 calories. Doing the same for an apple pie, you find that it has 372 calories.
To find the total calories in George's order, add: 260 + 320 = 580.

Zuzu's order has more calories. Therefore, subtract the number of calories in George's order from the number in Zuzu's order.

$$1,012 - 580 = 432$$

Zuzu's order has 432 more calories than George's order.

Step 4: Check your work.
First check the additions with subtractions.

Zuzu's order: 1,012 (total calories) – 372 (calories from the french fries) = 640 (calories from the Megaburger, or, alternately, 1,012 – 640 = 372 (calories from the french fries) Correct!

George's order: 580 (total calories) – 260 (calories from the hamburger) = 320 (calories from the apple pie), or, alternately, 580 – 320 = 260 (calories from the hamburger) Correct!

Next, check the subtraction with addition.

580 (George's total calories) + 432 (more calories in Zuzu's order) = 1,012 (total calories in Zuzu's order) Correct again!

EXAMPLE:
Rachel's family took a car trip from Boston to Cleveland. Hyun's family took a car trip from Cleveland to New York and then from New York to Boston. Whose family drove the greater distance? How many kilometers more?

Distance Chart (in kilometers)

| | Arrival City | | | |
Departure City	Boston, MA	New York, NY	Cleveland, OH	Los Angeles, CA
Boston, MA	—	350	1,035	5,040
New York, NY	350	—	810	4,685
Cleveland, OH	1,035	810	—	4,000
Los Angeles, CA	5,040	4,685	4,000	—

Step 2: Plan a strategy: Find the information.
What operations are suggested? addition to find the total kilometers that Hyun's family drove; subtraction to find the number of kilometers more that one family drove than the other family
Estimate the answer. Hyun's family drove about 100 kilometers more than Rachel's family (Hyun's family drove about 800 + 350 = 1,150 kilometers, while Rachel's family drove a little over 1,000 kilometers.)

Step 3: Do the plan.
You need to find the total distances traveled by both families and then compare them to see whose family drove the greater total distance. Then you can calculate how many more kilometers one family traveled than the other.
Look for words that may help you use the chart correctly. If Rachel's family went *from* Boston, it means that Boston was the starting point, or the point of departure. Going *to* Cleveland implies that Cleveland was the destination, or the point of arrival. Look under "Departure City" to find Boston. Put your finger or a pencil on Boston. Move across until you are in the column labeled "Cleveland, OH." Your finger should be on the number 1,035. It is 1,035 kilometers from Boston to Cleveland.

Do the same for Hyun's family. Hyun's family went *from* Cleveland (Cleveland is the departure city) *to* New York (New York is the first arrival city). Look under the

"Departure City" column for Cleveland. Move straight across to the column labeled "New York, NY." You should be on the number 810. Hyun's family traveled 810 kilometers from Cleveland to New York.

The family then left New York and traveled *to* Boston. Find New York under the column "Departure City," and then move straight across to the column labeled "Boston, MA." Are you now on the number 350? Good! You are doing great!

Now, let's finish the problem. Hyun's total distance is the distance traveled to New York (810 kilometers) combined with the distance traveled to Boston (350 kilometers). To find the total distance, add the two distances.

$$810 \text{ km} + 350 \text{ km} = 1,160 \text{ km}$$

Rachel's family traveled 1,035 kilometers and Hyun's family traveled 1,160 kilometers. Which family traveled the greater distance? The number 1,160 is greater than 1,035, so Hyun's family traveled the greater distance. How many more kilometers did Hyun's family travel than Rachel's family? Subtract to find how many more.

$$1,160 \text{ km} - 1,035 \text{ km} = 125 \text{ km}$$

Hyun's family traveled 125 kilometers more than Rachel's family.

Step 4: Check your work.
First, check to make sure you read the table correctly.

The distance from Boston to Cleveland is 1,035 kilometers.
The distance from Cleveland to New York is 810 kilometers.
The distance from New York to Boston is 350 kilometers.

Second, check the total distance that Hyun's family traveled. The total distance (1,160 kilometers) was found by addition. Check the addition by subtraction.

$$1{,}160 \text{ km} - 810 \text{ km} = 350 \text{ km, or} \qquad \checkmark$$

$$1{,}160 \text{ km} - 350 \text{ km} = 810 \text{ km} \qquad \checkmark$$

Last, check the 125-kilometer difference with addition.

$$125 \text{ km} + 1{,}035 \text{ km} = 1{,}160 \text{ km} \qquad \checkmark$$

Learning how to solve a problem
Takes time, but don't fret;
As we sharpen our word problem skills
The easier it will get!

BRAIN TICKLERS
Set # 5

For each word problem, plan a strategy by finding the needed information from the given display. Then solve the problem. Remember to check your work.

1. Jerold Jaunt will rent a compact car for a 10-day business trip. His company will pay the rental fee. After returning the business rental, he will rent a mid-size car for the next five days, when his family will join him on the trip.

   a. What is the cost of renting a luxury car for three days?

   b. What is the *minimum* cost for Jerold's company to rent a compact car for the first ten days?

c. What is the *minimum* cost for Jerold to rent a mid-size car for the additional five days?

Clem's Auto Rentals

Type	Daily Rate	3-day Rate	Weekly Rate (7 days)
Compact	$45	$120	$260
Mid-size	$55	$150	$345
Luxury	$75	$200	$500

2. Three friends were discussing the television shows that they watched on Monday night. Adam watched *Discover!*, *Family Fun*, and *LAPD Red*. Elana watched *Jackpot City* and *NYC 10021*. Arthur watched the *Financial News*, *Sports Spot*, *Music News*, and *Nightly News*.

   a. Which friend watched television for the greatest number of minutes?

   b. Another friend, Sophie, also watched television on Monday night. She watched only Channel 4 for a total of 90 minutes. Which combinations of television programs could she have watched?

Monday Night Television Programs

Channel	Time						
	7:00	7:30	8:00	8:30	9:00	9:30	10:00
4	Discover!	Science!	Music News		Nightly News		
5	Jackpot City	NYC 10021		Family Fun	LAPD Red		
7	Financial News	Sports Spot	Up to the Minute		Chicago City		

3. a. Jessica Journal delivers newspapers to families around her neighborhood. She delivers 42 newspapers Monday through Saturday and 50 newspapers on Sunday. Use the calendars below to find how many more newspapers Jessica delivered in January, 2000 than she delivered in February, 2000.

b. Zachary and Michaela both have part-time jobs at Stop and Save Supermarket. Zachary works five hours every fifth day, and Michaela works four hours every fourth day. Both worked on February 24, and each earned $7 per hour. How much money did each person earn during the months of January and February, 2000?

c. Who earned more money?

d. How much more were the earnings of one person than the earnings of the other?

Calendars

January 2000

S	M	T	W	Th	F	S
						1
2	3	4	5	6	7	8
9	10	11	12	13	14	15
16	17	18	19	20	21	22
23	24	25	26	27	28	29
30	31					

February 2000

S	M	T	W	Th	F	S
		1	2	3	4	5
6	7	8	9	10	11	12
13	14	15	16	17	18	19
20	21	22	23	24	25	26
27	28	29				

(Answers are on page 56.)

WORD PROBLEMS: STRATEGIES FOR SOLUTIONS

When you read a problem
And a strategy you can't find,
Make a list,
Guess and check,
Draw a picture,
Or find a pattern, if you don't mind.

When you're ready to wrestle with a word problem, you need a solution strategy. Different problems require different strategies. Some of these strategies are to make an organized list, to guess and check, to draw a picture or diagram, and to look for a pattern. But don't worry that you have to find the "one way" to solve a problem—you can often use more than one strategy to solve a given word problem.

Make an organized list

EXAMPLE:

Sahar, Phuong, and David bought a pack of six cookies to share with each other. If all of the cookies remain whole, in

how many ways can the cookies be distributed so that each boy receives at least one cookie?

Step 2: Plan a strategy: Make a list.

The phrase *share with* suggests division. However, the cookies are not necessarily shared equally. Therefore, you do not simply divide. Also, the question does not ask how many cookies each person gets, but in how many ways the cookies can be distributed. *In how many ways* means that you need to find all of the different possible ways, then count them. To do this it is wise to make an organized list of all of the possibilities.

Your list will need headings for who receives the cookies: Sahar, Phuong, and David. Under each name, put the number of cookies that person gets. Here are the heads for a list, which you can organize in the form of a chart.

Sahar   Phuong   David   Total Number of Cookies

You know that there are six cookies in all and that each person must receive at least one cookie. This means that you are looking for three numbers that add up to six, and that one of the numbers cannot be zero.
One way to start the organized list is to give Sahar one cookie. This leaves 6 − 1 = 5 cookies to distribute between Phuong and David.

Step 3: Do the plan.

Let's find all of the different ways to distribute five cookies between Phuong and David.

Sahar	Phuong	David	Total Number of Cookies
1	1	4	$1 + (1 + 4) = 1 + 5 = 6$
1	2	3	$1 + (2 + 3) = 1 + 5 = 6$
1	3	2	$1 + (3 + 2) = 1 + 5 = 6$
1	4	1	$1 + (4 + 1) = 1 + 5 = 6$
~~1~~	~~5~~	~~0~~	Oops! David needs to have a cookie. This way is not allowed.

Notice that giving two cookies to Phuong and three cookies to David is different from giving three cookies to Phuong and two cookies to David. Also, giving one cookie to Phuong and four cookies to David is different from giving four cookies to Phuong and one cookie to David.

So, you can see that there are four possible ways to distribute six cookies if Sahar receives only one cookie. Do you see how we've organized Phuong's number of cookies? We first gave him one cookie and then increased his number of cookies to two, then three, then four, until there were no more possibilities.

Now let's give Sahar two cookies. This will leave us with $6 - 2 = 4$ cookies to give to Phuong and David. Add these possibilities to your list.

Sahar	Phuong	David	Total Number of Cookies
1	1	4	
1	2	3	
1	3	2	
1	4	1	
2	1	3	$2 + (1 + 3) = 2 + 4 = 6$
2	2	2	$2 + (2 + 2) = 2 + 4 = 6$
2	3	1	$2 + (3 + 1) = 2 + 4 = 6$
~~2~~	~~4~~	~~0~~	Oops! David still needs to have a cookie! This way is not allowed.

There are three possible ways to give out six cookies if Sahar receives only two cookies.

Now continue your list by giving Sahar three cookies. This leaves $6 - 3 = 3$ cookies for the other two people.

3	1	2	$3 + (1 + 2) = 3 + 3 = 6$
3	2	1	$3 + (2 + 1) = 3 + 3 = 6$
~~3~~	~~3~~	~~0~~	Oops! David needs a cookie! This way is not allowed.

There are two possible ways to distribute six cookies if Sahar receives three cookies.

Now give Sahar four cookies. This leaves $6 - 4 = 2$ cookies for the other two.

4	1	1	$4 + (1 + 1) = 4 + 2 = 6$
~~4~~	~~2~~	~~0~~	Poor David! Where are his cookies? Cross out this way.

There is only one way to give out six cookies if Sahar receives four cookies.

So far we have $4 + 3 + 2 + 1 = 10$ ways to give out the six cookies.

Try giving Sahar five cookies. This leaves only one

cookie to be distributed between two people. Since you
can't cut up the cookies, this way will not work. Also,
giving all six of the cookies to Sahar will not work.

Making an organized list helps find the solution: There
are ten different ways to distribute six cookies among
three people if each person receives at least one cookie.

Step 4: Check your work.
If you study your list carefully you will see that you
have found all of the ways to distribute the cookies.

Sahar	Phuong	David	Total Number of Cookies
1	1	4	$1 + (1 + 4) = 1 + 5 = 6$
1	2	3	$1 + (2 + 3) = 1 + 5 = 6$
1	3	2	$1 + (3 + 2) = 1 + 5 = 6$
1	4	1	$1 + (4 + 1) = 1 + 5 = 6$
2	1	3	$2 + (1 + 3) = 2 + 4 = 6$
2	2	2	$2 + (2 + 2) = 2 + 4 = 6$
2	3	1	$2 + (3 + 1) = 2 + 4 = 6$
3	1	2	$3 + (1 + 2) = 3 + 3 = 6$
3	2	1	$3 + (2 + 1) = 3 + 3 = 6$
4	1	1	$4 + (1 + 1) = 4 + 2 = 6$

There are four different ways that Sahar can get one
cookie, three different ways he can get two cookies, two
different ways he can get three cookies, and one way he
can get four cookies. If you count the ways in the table that
Phuong (and also David) can get one, two, three, or four
cookies, you get the same number of ways as for Sahar.

Guess and check

Sometimes the easiest way to solve a word problem is to guess a
possible answer and then check to see if your guess works.
First, guess a number that might fit the conditions of the problem.

Second, check your guess. Does it satisfy the conditions of the problem?

If it checks, you are done! If it doesn't check, then guess again. Continue guessing until you have a guess that satisfies the conditions of the problem.

(NOTE: Keep track of your guesses. This will help you come closer to the solution.)

EXAMPLE:

Xeno bought a CD and a tape. Without the tax, the total cost was $20. The CD cost $10 more than the tape. Find the cost of each item.

Step 2: Plan a strategy: Guess and check.

Let's make a list of guesses that might work for Xeno's CD and tape.

Be sure to keep track of your guesses.

CD Cost	Tape Cost	Total (should equal $20)	Difference (should be $10)
$10	$10	$20 (10 + 10 = 20)	$0 ($10 − $10 = $0 more) Try another guess.
$11	$9	$20 (11 + 9 = 20)	$2 ($11 − $9 = $2 more) Try again!
$12	$8	$20 (12 + 8 = 20)	$4 ($12 − $8 = $4 more) Try again!
$13	$7	$20 (13 + 7 = 20)	$6 ($13 − $7 = $6) Getting closer!
$14	$6	$20 (14 + 6 = 20)	$8 ($14 − $6 = $8) Very close!
$15	$5	$20 (15 + 5 = 20)	$10 ($15 − $5 = $10) This is it!!!

The CD cost $15 and the tape cost $5.

MATH NOTE

Problems such as "Given the sum and difference of two numbers, what are the numbers?" appear in many algebra books. This type of problem can be solved using equations. We will talk about equations later in this book. For many problems, where the numbers are not too large, the guess and check plan works very well.

Draw a picture or diagram

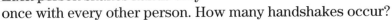

EXAMPLE:

There are four people at a party. Each person shakes hands only

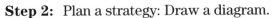

once with every other person. How many handshakes occur?

Step 2: Plan a strategy: Draw a diagram.
At first glance you might think that the answer is
4 people × 4 shakes or 16 handshakes. But wait, is this
logical? Would you shake your own hand? Maybe it is
4 people × 3 shakes per person or 12 shakes? Let's
make diagrams to solve the problem.

Step 3: Do the plan.
Let's name the four people A, B, C, and D. With diagrams we can show the handshakes.
Person A shakes hands with Persons B, C, and D, making three handshakes.

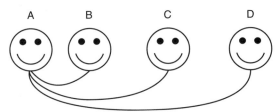

(NOTE: It takes two people to make one handshake. Thus,
when Person A shakes hands with Person B, Person B,
at the same time, shakes hands with Person A.)

Person B shakes hands with Persons C and D, making
two more handshakes. (Person B's handshake with
Person A has already been counted.)

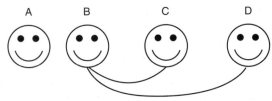

Person C shakes the hand of Person D, making one more handshake. (Person C's handshakes with Person A and Person B have already been counted.)

Everyone has already shaken the hand of Person D, so Person D need not shake any more hands.
There are 3 + 2 + 1 = 6 handshakes among the four people at the party.

Notice that you could also have solved this problem by making an organized list. Following is Step 3, using this alternative strategy.

Step 3: Do the plan.
List all the people with whom A shakes hands.
A with B   A with C   A with D = 3 handshakes

List the handshakes with B not already listed.
B with C   B with D   = 2 more handshakes

Do the same for C.
C with D   = 1 more handshake

Do the same for D.   0 more handshakes
                                       ─────────────
                                       6 handshakes

Again, there are 6 handshakes among the four people at the party.

Step 4: Check your work.
One way to check a result is to approach the problem from two or more different ways and arrive at the same solution. This was done in this example by using two different

strategies: making diagrams and making an organized list. By both methods, the same solution was obtained.

Look for a pattern

Sometimes in a problem
A pattern will step out;
It can make a big problem simpler,
There is no doubt!

The next example extends the previous example to a greater number of people by finding a pattern.

EXAMPLE:
There are nine people at a party. Each person shakes hands only once with every other person. How many handshakes occur?

Step 2: Plan a strategy: Look for a pattern.

Drawing a picture or making a list of all the handshakes would be very tiring. When you have a "big" word problem, it is often easier to simplify the problem, organize the simpler data in a table, and look for patterns that may develop.

Step 3: Do the plan.

For 3 people there are 3 handshakes $(2 + 1 = 3)$.
For 2 people, only one handshake will occur, and, simplest of all, with one person no handshakes will occur. This data has been entered in the table below along with the result from the previous Example. Notice the pattern of increases: first an increase of 1, then 2, and then 3. Use this pattern to predict the number of handshakes if there are 5 people at the party. Use your own method to check your guess.

Number of People	Number of Handshakes	
1	0 ←	+1
2	1 ←	+2
3	3 ←	+3
4	6 ←	+?
5	?	

You should find that there are 10 handshakes when there are 5 people at the party. Continue this pattern (or another pattern you may see) to extend the table for 6, 7, 8, and 9 people.

Number of People	Number of Handshakes
1	0
2	1
3	3
4	6
5	10
6	15
7	21
8	28
9	36

If there are 9 people at the party, 36 handshakes occur.

(Interesting fact: A man in India set the world record for handshaking in 1996. He shook 31,118 people's hands in eight hours! Imagine if all 31,118 people shook hands with each other!)

Step 4: Check your work.

The table was extended by continuing an arithmetic pattern. Therefore, you should go back to the table and check your additions.

EXAMPLE:

Lorna Leftfoot has started to walk for exercise. She walked 1 mile the first week, 5 miles the second week, 9 miles the third week, 13 miles the fourth week, and 17 miles the fifth week. If she continues to increase her walking distance this way, in which week will Lorna first walk 37 miles?

Step 2: Plan a strategy: Look for a pattern.

First make a table of the weeks and the number of miles walked each of the first four weeks. Then look for patterns.

Step 3: Do the plan.

Week	Miles Walked
1	1
2	5
3	9
4	13

How many more miles did Lorna walk in Week 2 than in week 1? (5 – 1 = 4 more miles)

How many more miles did Lorna walk in Week 3 than in Week 2? (9 – 5 = 4 more miles)

How many more miles did Lorna walk in Week 4 than in Week 3? (13 – 9 = 4 more miles)

If Lorna continues in this way, there will be 4 more miles from each week to the next.

Extend the table using the difference of four from one week to the next.

Week	Miles Walked
5	13 + 4 = 17
6	17 + 4 = 21
7	21 + 4 = 25
8	25 + 4 = 29
9	29 + 4 = 33
10	33 + 4 = 37!!!

Lorna will first walk 37 miles in the tenth week.

You may have realized that you didn't have to extend the table all the way up to the tenth week to solve the problem. Notice that the number of times that you add 4 miles is one less than the week number.

Week 2: $\underline{1} \times 4 =$ 4 additional miles

Week 3: $\underline{2} \times 4 =$ 8 additional miles

Week 4: $\underline{3} \times 4 = 12$ additional miles

In the week that Lorna first walks 37 miles, she has added 36 miles to the original distance of 1 mile. Because there are $36 \div 4 = 9$ groups of 4 in 36, it must be the tenth week, since the week number is one more than the number of groups of 4 miles added.

Step 4: Check your work.
Solving the problem by observing patterns in two different ways, and getting the same answer, provides a check on the solution.

Do you like one strategy better than another?
That's all right!
Draw a picture, make a list,
Or see a pattern or two;
Even if you guess and check,
Do the easiest one for you!

BRAIN TICKLERS
Set # 6

Solve each word problem by planning and carrying out a strategy. Remember to check your work. There is a letter at the end of each problem. Place this letter on the line corresponding to the numerical answer to the problem in the answer code at the end of this problem set. When you complete the answer code, you will find the answer to this question:

What famous candy did Clarence Crane first produce in 1912?

1. In a round-robin softball tournament, every one of the ten teams in the league must play every other team once and only once. Altogether, how many games will be played? (R)

2. In a barnyard there is a total of eight chickens and pigs. There are 26 legs altogether. How many chickens are in the barnyard? (S)

3. In the sport of archery, the target has five colored rings: gold, red, blue, black, and white. How many different ways can four arrows hit the target if *each* arrow hits *either* the gold, blue, or black ring? (Remember that more than one arrow may hit the same-colored ring and that all of the rings need not be hit.) (L)

4. Blair, Sam, and Jaclyn each invited one friend to a picnic. Each of their friends invited two friends to the picnic, and each of these friends invited three other friends to the picnic. If everyone invited came, how many people ended up at the picnic? (E)

5. Bart, Burt, Bella, and Bonnie Baseman are siblings who share two tickets to the Planets baseball games. If the tickets are for one game per week, for how many weeks can the Basemans attend the games without repeating the same combination of siblings? (V)

6. The Thurston School is building a new wing as more families are moving into town. The following numbers of families have moved into town over the past four years: 3 more families the first year, then 8 more families the following year, then 13 more families the next year, and 18 more families the past year. If this pattern continues, in how many more years will a total of 43 new families have moved into town? (F)

7. Krystal works on Mondays, Wednesdays, and Saturdays at Discount City. She works for a total of 12 hours, and she can split up her hours over the three days in any way that she chooses as long as she works a minimum of three hours per day. If she cannot work a fractional part of an hour, in how many ways can she split up her hours over the three days? (A)

8. The total cost of a movie ticket and popcorn is $11. The ticket costs $5 more than the popcorn. Find the cost of the movie ticket. (S)

9. Reed Bocker has started to jog. He jogged one mile the first week, two miles the second week, four miles the third week, seven miles the fourth week, and eleven miles the fifth week. If he continues increasing his distance in this pattern, how many miles will he jog in the tenth week? (I)

10. The sum of the digits of a two-digit number is nine. The difference between the two digits is three, and the tens digit is larger than the units digit. What is the number? (E)

Answer Code

Clarence Crane produced:

$\overline{}$	$\overline{}$	$\overline{}$	$\overline{}$		$\overline{}$	$\overline{}$	$\overline{}$	$\overline{}$	$\overline{}$	$\overline{}$
15	46	5	30		8	10	6	63	45	3

(Answers are on page 59.)

BRAIN TICKLERS—THE ANSWERS

Set # 4, page 27

1. Estimate: more than 300 cards (250 + 70 > 300)
   Add to find the number of cards in Felicia's collection.

$$75 + 248 = 323 \text{ cards}$$

There are 323 cards in Felicia's collection.

Check addition with subtraction.

$$323 - 75 = 248 \text{ cards for Ben}$$

CROSS OUT 323 IN THE SOLUTION BOX.

2. Estimate: about 30 (about 50 players altogether on the Quails and Ducks teams, and 18 fewer on the Buzzards team)
   First, add to find the total number of players on the Quails and Ducks teams.

$$24 + 27 = 51 \text{ players}$$

Then, subtract to find the number of players on the Buzzards team.

$$51 - 18 = 33 \text{ players}$$

There are 33 players on the Buzzards team.

Check: Work backwards.
First, check subtraction with addition.

$$33 + 18 = 51 \text{ players altogether on the Quails and Ducks teams}$$

Then, check addition with subtraction.

$$51 - 27 = 24 \text{ players on the Quails team } \checkmark$$

CROSS OUT 33 IN THE SOLUTION BOX.

3. Estimate: difficult without working problem
   First, multiply to find the total that Jacob collected.

12 sponsors × $4 donated per mile = $48 collected per mile

$48 × 8 miles = $384 total collected by Jacob

Second, multiply to find the total that Lisa collected.

18 sponsors × $2 donated per mile = $36 collected per mile

$36 × 8 miles = $288 total collected by Lisa

Then, subtract to find how much more Jacob collected than Lisa.

$384 − $288 = $96

Jacob collected $96 more than Lisa.

Check: Work backwards.
First, check subtraction with addition.

$96 + $288 = $384 collected by Jacob

Then check each multiplication by division.

$288 ÷ 8 = $36 for each mile Lisa walked

$36 ÷ $2 = 18 sponsors for Lisa ✓

$384 ÷ 8 = $48 for each mile Jacob walked

$48 ÷ $4 = 12 sponsors for Jacob ✓

CROSS OUT 96 IN THE SOLUTION BOX.

4. Estimate: about $800 (1,000 + 300 − 500)
   First, add the $250 deposit to the initial $1,000.

$1,000 + $250 = $1,250

Then, subtract the $540 withdrawal.

$1,250 − $540 = $710

The balance after the deposit and withdrawal is $710.

Check: Work backwards.
First, check the subtraction with addition.

$710 + $540 = $1,250

Then, check the addition with subtraction.

$1,250 − $250 = $1,000, the starting balance ✓

CROSS OUT 710 IN THE SOLUTION BOX.

5. Estimate: about 3,000 (Each cubit is less than 2 feet long; 5,280 feet divided into 2-foot lengths is over 2,500.)
Draw a picture to help you solve the problem.

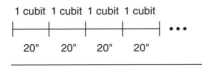

5,280 feet

First, multiply to change feet to inches.

5,280 ft × 12 in. per ft = 63,360 in.

Then, divide to find how many 20-inch units (cubits) are in 63,360 inches.

63,360 in. ÷ 20 in. per cubit = 3,168 cubits

3,168 of Josh's cubits laid end to end would cover a length of one mile.

Check: Work backwards.
First, check division with multiplication.

3,168 bills × 20 in. per cubit = 63,360 in.

Then, check multiplication with division.

63,360 ÷ 12 in. per ft = 5,280 ft

CROSS OUT 3,168 IN THE SOLUTION BOX.

6. Estimate: more than 80 (Currently, Kareem has scored more points below 80 than above 80.)
Add to find how many points Kareem has scored so far.

73 + 85 + 81 + 71 = 310 points

Subtract to find how many points Kareem still needs to have a total of 400.

$$400 - 310 = 90 \text{ points}$$

Kareem needs 90 points on the fifth test to have an average of 80.

Check: Work backwards.
First, check subtraction with addition.

$$90 + 310 = 400 \text{ points}$$

Then check your original addition to make sure the scores add to 310 points.

CROSS OUT 90 IN THE SOLUTION BOX.

7. Estimate: more than $20 (4 hr × $5/hr = $20)
   First, find how long the car was parked at the stated rates.

   For the first hour (from 1:00 to 2:00): $6/hr

   For the next three hours
   (2:00 to 3:00, 3:00 to 4:00, 4:00 to 5:00): $5/hr

   For the last half hour (5:00 to 5:30): also $5/hr,
   since parts of an hour are charged at the whole
   hourly rate of $5/hr

   Then multiply each number of hours by the correct price per hour.

   $$1 \text{ hr} \times \$6/\text{hr} = \$6$$

   $$4 \text{ hr} \times \$5/\text{hr} = \$20$$

   Last, add to find the total cost of parking the car.

   $$\$6 + \$20 = \$26$$

   It costs $26 to park the car for the hours given.

   Check: Work backwards.
   First, check addition with subtraction.

$$\$26 - \$6 = \$20 \text{ or } \$26 - \$20 = \$6$$

Then check the multiplications by divisions.

$$\$20 \div \$5/hr = 4 \text{ hr}$$

$$\$6 \div \$6/hr = 1 \text{ hr}$$

CROSS OUT 26 IN THE SOLUTION BOX.

8. Estimate: more than $300 ($150 for 10 months; $300 for 20 months; 20 months is less than 2 years.)
   First, multiply to change 2 years to months.

$$2 \text{ yr} \times 12 \text{ mo/yr} = 24 \text{ mo}$$

Then, multiply to find the amount paid in installments in 24 months.

$$\$15/mo \times 24 \text{ mo} = \$360$$

Then, add to find the total cost, including the $25 deposit.

$$\$360 + \$25 = \$385$$

The total cost of the stereo system was $385.

Check: Work backwards.
First, check addition with subtraction.

$$\$385 - \$25 = \$360$$

Then check the multiplications by divisions.

$$\$360 \div \$15/mo = 24 \text{ mo}$$

$$24 \text{ mo} \div 12 \text{ mo/yr} = 2 \text{ yr} \checkmark$$

CROSS OUT 385 IN THE SOLUTION BOX.

9. Estimate: about 60 boxes (More than $10 per box for note-paper means more than $550 was collected for the notepaper; more than $1,200 collected altogether means roughly $500 or $600 was collected for the note cards; at $8 per box, more than 60 would be sold.)

First, multiply to find the total collected from the sale of notepaper.

$$\$12 \text{ per box} \times 55 \text{ boxes} = \$660$$

Second, subtract to find the amount collected for the sale of notecards.

$$\$1,236 - \$660 = \$576$$

Last, divide to find how many boxes of note cards were sold.

$$\$576 \div \$8 \text{ per box} = 72 \text{ boxes}$$

72 boxes of note cards were sold.

Check: Work backwards.
First, check division with multiplication.

$$72 \text{ boxes} \times \$8 \text{ per box} = \$576$$

Then, check subtraction with addition.

$$\$576 + \$660 = \$1,236$$

Last, check multiplication with division.

$$\$660 \div 55 \text{ boxes} = \$12 \text{ per box for the notepaper}$$

CROSS OUT 72 IN THE SOLUTION BOX.

10. Estimate: about $700 (The car cost more than $1,800 with repairs; the profit from the sale of the car was roughly $2,800, which was divided equally among the four friends. First, add to find the total spent on the car.

$$\$1,250 + \$575 = \$1,825$$

Then, subtract to find the profit made on the sale of the car.

$$\$4,605 - \$1,825 = \$2,780$$

Last, divide to find how much each friend received.

$$\$2,780 \div 4 = \$695$$

Each friend received $695.

Check: Work backwards.
First, check division with multiplication.

$$\$695 \times 4 = \$2,780 \text{ total profit}$$

Then, check subtraction with addition.

$$\$2,780 + \$1,825 = \$4,605$$

Last, check addition with subtraction.

$$\$1,825 - \$575 = \$1,250$$

CROSS OUT 695 IN THE SOLUTION BOX.

The only number left in the solution box is 300. For the movie *The Wizard of Oz*, 300 pairs of shoes were dyed emerald green for the final procession.

Set # 5, page 34

1. a. $200 (Read from the table.)
    b. To determine the *mini- mum* cost for a 10-day rental of a compact car, use the information in the  chart. Under the column labeled "Type" find "Compact" and read along this line. A 10-day rental of a compact car can consist of 10 days at the daily rate of $45 (10 × $45 = $450), or three 3-day periods plus one more day (3 × $120 + $45 = $360 + $45 = $405), or as a full week (7 days) plus a 3-day period ($260 + $120 = $380). Since $380 is less than either of the other two costs, the minimum cost is $380.
    c. To determine the *minimum* cost for a 5-day rental of a mid-size car, again use the information in the chart. Under the column labeled "Type" find "Mid-size" and read along this line. A 5-day rental of a mid-size car can consist of five days at the daily rate of $55 (5 × $55 = $275), or as one 3-day period plus two more days [$150 + (2 × $55) = $150 + $110 = $260]. Since $260 is less than $275, $260 is the minimum cost.

2. a. To find the time that Adam watched television, use the television program to find the information.
    Adam watched the following programs:

    > Discovery! from 7 to 7:30, or 30 minutes
    > Family Fun from 8:30 to 9:00, or 30 minutes
    > LAPD Red from 9:00 to 10:00, or 60 minutes

    Adam watched television for a total of 120 minutes.
    Now find the time that Elana watched television.
    Elana watched the following programs:

    > Jackpot City from 7 to 7:30, or 30 minutes
    > NYC 10021 from 7:30 to 8:30, or 60 minutes

    Elana watched television for a total of 90 minutes.
    Last, find the time that Arthur watched television.
    Arthur watched the following programs:

    > Financial News from 7 to 7:30, or 30 minutes
    > Sports Spot from 7:30 to 8:00, or 30 minutes
    > Music News from 8:00 to 9:00, or 60 minutes
    > Nightly News from 9:00 to 10:00, or 60 minutes

    Arthur watched television for a total of 180 minutes.
    Arthur watched television for the greatest number of minutes.
    b. Find "4" in the column labeled "Channel" and read across.
    Sophie could have watched the following on Channel 4 for a total of 90 minutes:

    > Discover! (30 minutes) and Music News (60 minutes)
    > Science! (30 minutes) and Music News (60 minutes)
    > Discover! (30 minutes) and Nightly News (60 minutes)
    > Science! (30 minutes) and Nightly News (60 minutes)

3. a. First find the number of daily newspapers Jessica delivers in January.

    > There are six days from Monday through Saturday. There are four Monday through Saturday intervals and one more Saturday and one more Monday in January, for a total of (4 × 6) + 2 = 26 days with daily newspapers delivered. That is a total of 26 days × 42 newspapers per day, or 1,092 daily newspapers delivered in January.

Now find the number of Sunday newspapers delivered in January.

> There are five Sundays in January, for a total of $5 \times 50 =$ 250 Sunday newspapers delivered.

Add to find the total number of newspapers that Jessica delivered in January.

> $1,092 + 250 = 1,342$ newspapers delivered in January

Follow the same steps to find the number of newspapers Jessica delivered in February.

> There are three Monday through Saturday intervals of six days and five extra days in the first week and two extra days in the last week in February for a total of $(3 \times 6) + 5 + 2 = 25$ days with daily newspapers delivered. That is a total of 25 days $\times$ 42 newspapers per day, or 1,050 daily newspapers delivered in February.

Now find the number of Sunday newspapers delivered in February.

> There are four Sundays in February, for a total of $4 \times 50 =$ 200 Sunday newspapers delivered.

Add to find the total number of newspapers that Jessica delivered in February.

> $1,050 + 200 = 1,250$ newspapers delivered in February

Last, subtract to find how many more newspapers Jessica delivered in January than in February.

$1,342 - 1,250 = 92$ more newspapers in January than February

b. First, find how many days Zachary worked.
   Put your pencil on February 24 (one day of work).
   Count ahead five days to February 29 for another day of work. Now, go back to February 24. Count back by fives to find Zachary's other days of work: February 19, 14, 9, and 4, and January 30, 25, 20, 15, 10, and 5, for another 10 days. Zachary worked 5 hours each of the $1 + 1 + 10 = 12$ days:

5 hours per day × 12 days = 60 hours

Zachary earned $7 per hour:

$7 × 60 hours = $420

Now find the days that Michaela worked.
Go back to February 24 (one day). Count ahead four days to February 28 for another day of work. Now, go back to February 24. Count back by fours to find Michaela's other days of work: February 20, 16, 12, 8, and 4, and January 31, 27, 23, 19, 15, 11, 7, and 3, for another 13 days. Michaela worked 4 hours each of the 1 + 1 + 13 = 15 days:

4 hours per day × 15 days = 60 hours

Michaela also earned $7 per hour:

$7 × 60 hours = $420

c. Both Zachary and Michaela earned $420.
d. They both earned the same amount.

Set # 6, page 48

1. First, simplify the problem by considering the number of games if there were 2, then 3, 4, and 5 teams. Use your own method for determining the number of games for these small numbers of teams. You might use drawings, or name teams and list the pairs of teams.
   Organize your data in a table. Look for patterns and extend the table.
   Your organized and extended table should look something like this:

Number of Teams	2	3	4	5	6	7	8	9	10
Number of Games	1	3	6	10	15	21	28	36	45

There are 45 games played when there are 10 teams.
Put an R on the line above 45.

2. Guess and check with a list of guesses. (If you know how to use an equation you may do so. Equations will be explained in Chapter Eight.)
   Guess two numbers whose sum equals eight.

Number of Chickens (2 legs)	Number of Pigs (4 legs)	Total Number of Animals (a sum of 8)	Total Number of Legs (26 in all)
1	7	$1 + 7 = 8$	$(1 \times 2) + (7 \times 4) = 30$
2	6	$2 + 6 = 8$	$(2 \times 2) + (6 \times 4) = 28$
3	5	$3 + 5 = 8$	$(3 \times 2) + (5 \times 4) = 26!$

There are three chickens in the barnyard.
Put an S on the line above 3.

3. Make an organized list with the colored rings. In each row the sum will be four for the four arrows. Remember that a ring may be hit more than once, and that all the colors need not be hit in a certain round of four arrows.

Gold	Blue	Black
4	—	—
3	1	—
3	—	1
2	2	—
2	1	1
2	—	—
1	3	—
1	2	1
1	1	2
1	—	3
—	4	—
—	3	1
—	2	2
—	1	3
—	—	4

There are 15 ways to hit the target.
Put an L on the line above 15.

4. Draw a picture ("x" represents one person). Here is one type of picture:

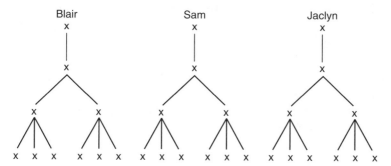

Count the x's for a total of 30 people. This diagram is called a *tree diagram*.
Put an E on the line above 30.

5. Make a list of combinations of two siblings. (Remember that the combination Bart-Burt is the same as the combination Burt-Bart, and so on.)

    Bart-Burt, Bart-Bella, Bart-Bonnie
    Burt-Bella, Burt-Bonnie
    Bella-Bonnie

    They can use the tickets for six weeks without repeating a combination of siblings.
    Put a V on the line above 6.

6. List the numbers to recognize a pattern.

    3, 8, 13, 18

    3 + **5** = 8; 8 + **5** = 13; and 13 + **5** = 18

    Notice that you are adding 5 each time! That means that 5 more families move in each year than the year before. If this pattern of adding 5 continues, you can extend the list until you reach a total of 43.

    18 + 5 = 23; 23 + 5 = 28; 28 + 5 = 33; 33 + 5 = 38; 38 + 5 = 43!

    In five more years (23, 28, 33, 38, 43), there will be 43 new families.

Put an F on the line above 5.

7. Make an organized list with the days Krystal works and the number of hours she works each day (at least three each day for a total of 12 hours).

Begin by listing all of the ways that Krystal can work 3 hours on Monday. (Subtract Monday's number from 12 to find the remaining number of hours needed: $12 - 3 = 9$ left. Find two numbers whose sum is 9 for the other two days. Remember that each number must be at least 3.)

Monday	Wednesday	Saturday	Total Hours
3	3	6	12
3	4	5	12
3	5	4	12
3	6	3	12
✗	✗	✗	Oops! Not allowed. Krystal needs to work at least 3 hours per day.

Next, list all of the ways that Krystal can work 4 hours, then 5 hours, and then 6 hours on Monday. (Find two numbers whose sum is the remaining number of hours for the other two days.)

4	3	5	12
4	4	4	12
4	5	3	12
5	3	4	12
5	4	3	12
6	3	3	12

There are ten ways to form Krystal's schedule.
Put an A on the line above 10.

8. Guess and check by finding two numbers whose sum is 11. It is reasonable to use a larger number for the ticket.

Movie Ticket ($)	Popcorn ($)	Total Cost ($)	Difference ($5 more for the Ticket)
6	5	11	6 − 5 = 1 Too small!
7	4	11	7 − 4 = 3 Closer!
8	3	11	8 − 3 = 5 Yippee!

The movie ticket costs $8.
Put an S on the line above 8.

9. Organize the given information in a table. Look for a pattern, and continue the pattern until you get to the tenth week.

Week	Miles Jogged	
1	1	
2	2	(1 + 1 = 2)
3	4	(2 + 2 = 4)
4	7	(4 + 3 = 7)
5	11	(7 + 4 = 11)

Pattern: From Week 1 to Week 2, add 1. From Week 2 to Week 3, add 2. From Week 3 to Week 4, add 3, and from Week 4 to Week 5, add 4. Look! The pattern is +1, +2, +3, +4. For the next week, add 5, then add 6, and so on. Continue until the tenth week.

Week	Miles Jogged	
6	16	(11 + 5)
7	22	(16 + 6)
8	29	(22 + 7)
9	37	(29 + 8)
10	46	(37 + 9)

Reed will jog 46 miles in the tenth week.
Put an I on the line above 46.

10. Guess and check two digits whose sum is 9. Subtract the digits to see if the difference is 3. The tens digit needs to be greater than the units digit.

Guess	Sum of Digits	Difference of Digits (3)
54	9	5 – 1 = 4 Nope!
63	9	6 – 3 = 3 Yeah!

The number is 63.
Put an E on the line above 63.

Answer Code

L	I	F	E		S	A	V	E	R	S
15	46	5	30		8	10	6	63	45	3

Clarence Crane produced Life Savers!

CHAPTER THREE

Daffy Decimals and Fun Fractions

A word problem with decimals
Can be solved with ease:
Find a strategy, do the plan,
And check it, if you please.

Decimal word problems

To solve a word problem with decimals, use the same strategies you have used to solve problems with whole numbers. Remember, before you can plan a strategy, you must understand the problem. Don't forget to answer these questions: What is the problem about? What are you being asked to find? Is there enough information? The key words you use to decide on the operation or operations to use are the same for decimal word problems as for whole number word problems. You may also have to decide if a picture, a list, a pattern, or a guess and check strategy will help you solve the problem.

MATH NOTE

Integers are also decimal numbers. If a decimal point is not visible, place it to the right of the units digit. For example, 52 is equal to 52., or 52.0.

DECIMAL OPERATIONS

Following are some examples preceded by sample calculations that will give you hints for performing operations with decimals as you solve word problems.

Hints for multiplying decimals

To multiply 0.02 by 52, first write the multiplication as if the numbers were whole numbers. (It is *not* necessary to line up the decimal points.)

$$\begin{array}{r} 0.02 \\ \times\ 52. \\ \hline \end{array}$$

Then, multiply the factors to find the product. Ignore the decimal points for now.

$$
\begin{array}{r}
0.02 \\
\times \quad 52. \\
\hline
4 \\
100 \\
\hline
104
\end{array}
$$

To place the decimal point in the answer, follow these steps:

From the right side of the right-most non-zero digit of each factor, count the number of spaces you need to hop back until you reach the decimal point.

$$0 . 0\ 2 \text{ (two spaces back)}$$

$$5\ 2. \text{ (zero spaces back)}$$

Total the spaces back that you hopped.

$$2 + 0 = 2 \text{ spaces hopped back}$$

Now, go to the product. From the right of the units digit, hop back the total number of spaces hopped from the two factors. This is where to place the decimal point in the product.

$$1\ 0\ 4$$

Decimal point goes here.

Moving the decimal point two places to the left gives a final product of 1.04.

Hints for dividing decimals

First write the division as if the numbers were whole numbers.

$$8.50\overline{)1,275}$$

If there isn't a decimal point shown in the divisor, place the decimal point to the right of the units digit. If there is a decimal point in the divisor, hop it to the right until it is after the units digit.

$$8.50.\overline{)1{,}275}$$

Decimal point goes here.

Count how many spaces were hopped. Go to the dividend (1,275) and move its decimal point over that many places. (1,275. will become 127,500 because you need to include two zeros as place holders.)

$$850.\overline{)1{,}27500.}$$

Put your pencil on the new decimal point in the dividend. Move straight up and place a point above the division sign. This is where the decimal point will be in the quotient.

$$850.\overline{)127500.}$$

Now divide as you would with whole numbers.

$$
\begin{array}{r}
150. \\
850.\overline{)127500.} \\
\underline{850} \\
4250 \\
\underline{4250} \\
0
\end{array}
$$

(If the answer is not a whole number, that is, there is a remainder, place zeros after the last numeral in the dividend. Bring as many zeros "down" as you need to find the answer.)

EXAMPLE:
Fingernails grow an average of 0.02 inch per week. If you never cut or broke your thumbnail for one year, how many inches long would it be? (Remember that one year equals 52 weeks.)

Step 2: Plan a strategy.

What operation is suggested? multiplication to find the total length

Estimate the answer. about 1 inch (In 100 weeks the nail would grow 2 inches, and in half that time it would grow about 1 inch.)

Step 3: Do the plan.

Multiply to find the total length.

$$
\begin{array}{r}
0.02 \text{ in./wk} \\
\times\ \ 52. \text{ wk} \\
\hline
1.04 \text{ in.}
\end{array}
$$

Your thumbnail would be 1.04 inches long.

Step 4: Check your work.

Check your multiplication with division.

$$
\begin{array}{r}
52. \checkmark \\
0.02.\overline{)1.04.} \\
\underline{10} \\
4 \\
\underline{4} \\
0
\end{array}
$$

EXAMPLE:

Movie Time Cinema charges $8.50 per person aged 13 or older and $4.25 per child under age 13. Friday night, Movie Time took in a total of $850 in children's tickets and $1,275 in tickets purchased by people aged 13 or older. How many more children's tickets were sold than adult tickets?

Step 2: Plan a strategy.

What operations are suggested? division to find the number of children's tickets sold; division to find the number of adult tickets sold (In each case, you know the total amount and the cost of each ticket.); then subtraction to find how many more children's tickets were sold than adult tickets

Estimate the answer. fewer than 100 more children's
tickets (About 800 ÷ 4 = 200 children's tickets were
sold, and more than 1000 ÷ 10 = 100 adult tickets were
sold.)

Step 3: Do the plan.
First, divide the total value of adult tickets sold by the
price of each adult ticket.
After moving the decimal points, divide as you would
with whole numbers.

$$8.50.\overline{)1275.00.} = 150.$$

There were 150 adult tickets sold.
Next, divide the total value of children's tickets sold by
the price of each ticket.

$$4.25.\overline{)850.00.} = 200.$$

There were 200 children's tickets sold.
Last, to find how many more children's tickets were
sold than adult tickets, subtract the number of adult
tickets sold from the number of children's tickets sold.

$$200 - 150 = 50$$

There were 50 more children's tickets sold.

Step 4: Check your work.
Work backwards.
First, check the subtraction with addition.

$$50 + 150 = 200 \text{ children's tickets sold} \quad ✓$$

Next, check your divisions with multiplications.

200 children's tickets sold × $4.25 per ticket = $850 ✓

150 adult tickets sold × $8.50 per ticket = $1,275 ✓

Hints for adding decimals

To add 2.75 and 1.30, first line up the decimal points. Then bring down the decimal point, and add as you would for whole numbers.

$$
\begin{array}{r}
2.75 \\
+\ 1.30 \\
\hline
4.05
\end{array}
$$

Hints for subtracting decimals

As in adding decimals, line up the decimal points and then bring down the decimal point. Subtract as you would for whole numbers.

$$
\begin{array}{r}
4.05 \\
-\ 1.30 \\
\hline
2.75
\end{array}
$$

EXAMPLE:
Luke bought popcorn for $2.75 and a soda for $1.30. Leia bought candy for $3.15 and lemonade for $1.25. Who spent more money on refreshments, and how much more was spent?

Step 2: Plan a strategy.
What operations are suggested? addition to find the total amount that Luke spent; addition to find the total amount that Leia spent; then subtraction to find the difference between the two amounts
Estimate the answer. Leia spent less than $1 more than Luke. (Both spent more than $4 but less than $5.)

Step 3: Do the plan.
First find Luke's total. Add the prices of the popcorn and soda.

$$
\begin{array}{r}
\$2.75 \\
+\ \$1.30 \\
\hline
\$4.05
\end{array}
$$

Luke spent $4.05 for refreshments.
Next find Leia's total. Add the prices of the candy and
lemonade.

$$\begin{array}{r} \$3.15 \\ + \ \$1.25 \\ \hline \$4.40 \end{array}$$

Leia spent $4.40 for refreshments.
Because $4.40 is greater than $4.05, Leia spent more
money.
Subtract to find how much more Leia spent.

$$\begin{array}{r} \$4.40 \\ - \ \$4.05 \\ \hline \$ \ .35 \end{array}$$

Leia spent $0.35 more than Luke.

Step 4: Check your work.
First check the subtraction with addition.

$$\begin{array}{r} \$0.35 \\ + \ \$4.05 \\ \hline \$4.40 \ \checkmark \end{array}$$

Then check the additions with subtractions.

$$\begin{array}{r} \$4.40 \\ - \ \$1.25 \\ \hline \$3.15 \end{array} \quad \text{or} \quad \begin{array}{r} \$4.40 \\ - \ \$3.15 \\ \hline \$1.25 \ \checkmark \end{array}$$

and

$$\begin{array}{r} \$4.05 \\ - \ \$1.30 \\ \hline \$2.75 \end{array} \quad \text{or} \quad \begin{array}{r} \$4.05 \\ - \ \$2.75 \\ \hline \$1.30 \ \checkmark \end{array}$$

BRAIN TICKLERS
Set # 7

Solve each word problem by reading it care-fully and thoroughly, planning a strategy, esti-mating the answer, doing the plan, and then checking your work. Then in your own words, list the steps you used to solve the problem.

The solution box following the problems con-tains the answers to the problems. As you an-swer each problem, cross out the answer in the solution box. Use the number that is not crossed out to answer the following question:

What is the weight, in tons, of the heaviest ice cream sundae ever made?

1. Victor's Video Arcade pays its employees $7.95 per hour for weekday hours and $9.25 per hour for weekend hours. How much will an employee earn for working six consecutive days for seven hours per day if the employee starts working on a Monday?

2. The diameter of an Oreo cookie measures 1.75 inches. To the nearest whole number, how many Oreos laid end to end would form a line one yard long?

3. A calligrapher charges $0.25 for every ten words printed and $2.25 per sheet of embossed paper. How much would the cal-ligrapher charge to print 300 words that require a total of eight sheets of paper?

4. The Boston Marathon is an annual road race of 26.2 miles. If one mile is about 1.6 kilometers, about how many kilometers long is the Boston Marathon?

5. Joshua is 181.0 centimeters tall. Rachel is 17.6 centimeters shorter than Joshua. Corbin is 15.2 centimeters taller than

Rachel. Janet is 0.4 centimeters taller than Corbin. How many centimeters tall is Janet?

6. Samuel Spender had $218.35 in his savings account. One week he made a withdrawal of $174.50, the next week he deposited a check for $34.99, and the next week he withdrew $18.25 to purchase a book. How much money remained in Samuel's account?

7. A phone call from Main City to Central Town costs $0.35 for the first three minutes and $0.25 for each additional minute. How many minutes long would a call be that costs a total of $1.85?

8. Crunchy candy bars cost $0.45 each and Wonder candy bars cost $0.55 each. Candace bought nine candy bars for a total of $4.45. How many Crunchy candy bars did she buy?

9. The hairs on your head grow at an average rate of 0.013 inch per day. To the nearest hundredth, how many inches long would your hair be if you did not cut your hair for three non-leap years?

10. Reggie brought seven liters of juice to the class picnic, and Sarah brought three gallons of juice. Reggie said that he brought more juice because seven is greater than three. If one liter is equal to 0.26 gallons, see if Reggie is correct. How many more gallons of juice did one person bring than the other?

Solution Box

9		179.0
41.92	343	5
14.24	60.59	22.59
1.18	25.50	21

(Answers are on page 93.)

Now it's time to search and solve
For data in a display.
Decimals appear everywhere,
So let's solve for an answer today.

USING FACTS IN A DISPLAY

You can find decimals in many types of displays: banking forms, catalogs, and mileage charts, to name a few. Use the same strategies that you used in Chapter 2 to solve word problems involving displays with decimals.

BRAIN TICKLERS
Set # 8

Solve each word problem, planning and carrying out a strategy by searching the display for the correct data. Remember to check your work.

1. Use the tables below to answer the questions following.

House and Home Catalog Prices	
Item	**Cost**
CD rack	$49.75
Magazine holder	$15.95
Pasta set	$35.49
Picnic set	$24.55
Mouse pad	$7.89
Disk caddy	$14.29

Shipping Charges	
Total Amount	**Shipping Charges**
Up to $15.99	$4.95
$16.00 to $30.99	$5.95
$31.00 to $50.99	$7.95
$51.00 to $75.99	$10.95
$76.00 to $99.99	$11.95
$100.00 or more	$15.00

a. Paul Placeman received $100 as a gift from his Aunt Bertha. He wanted to order two magazine holders, one mouse pad, and one CD rack. Including the shipping charges, will Paul's gift be enough for the purchase? If so, how much money will remain? If not, how much more will he need?

b. Gila Guesster ordered two each of two items from *House and Home*. Including the shipping charges, her total was $111.51. Which items did she order?

2. Michelle forgot to put in the full entries into her checking account record. Use Michelle's entries below to answer the questions following.

Check#	Date	Transaction	Payment/Withdrawal	Deposit	Balance
					$329.42
124	6/29	City Cable	???	—	$293.59
—	7/2	—	—	$145.85	???
—	7/5	withdrawal	???	—	$373.94

a. Find the amount of the check payment to City Cable on June 29.

b. After the July 2 deposit, what was the new balance?

c. Find the amount of the withdrawal made on July 5.

3. The chart below gives the winners of the men's shot put for several Olympic games. Use the chart to answer the questions following.

Year	Winner	Country	Distance of Throw (meters)
1980	Vladimir Kyselov	USSR	21.35
1984	Alessandro Andrei	Italy	21.26
1988	Ulf Timmerman	East Germany	22.47
1992	Michael Tulce	United States	21.70
1996	Randy Barnes	United States	21.62

a. Order the men's shot put winners from greatest distance thrown to least distance thrown.

b. Find the difference between the greatest shot put distance and the least distance.

c. Find the sets of consecutive Olympic years in which the distances increased from one Olympics to the next. Between which two years was the increase the greatest?

(Answers are on page 100.)

FRACTION OPERATIONS

Fractions are what we'll tackle next
In sales, in stocks, and more.
Fractions are a part of life
So let's see what we find in store.

Following are some examples preceded by sample calculations that will give you hints for performing operations with fractions and mixed numbers as you solve word problems.

Hints for multiplying a fraction by a whole number

To find $\frac{1}{3}$ of 48, first multiply 48 by the *numerator* (top part) of the fraction.

$$48 \times 1 = 48$$

Next, divide the result by the *denominator* (bottom part) of the fraction.

$$48 \div 3 = 16$$

So, $\frac{1}{3}$ of 48 is 16.

MATH NOTE

Notice that multiplying by $\frac{1}{3}$ is the same as dividing by 3.

Below are some other easy ways to multiply with *unit fractions* (fractions with a numerator of 1).

Multiplying by:	is the same as	Dividing by:
$\dfrac{1}{2}$		2
$\dfrac{1}{4}$		4
$\dfrac{1}{5}$		5

Other unit fractions follow the same pattern.

Remember that 2 can also be written as $\dfrac{2}{1}$. Multiplying by $\dfrac{1}{2}$ is the same as dividing by its *reciprocal*, $\dfrac{2}{1}$; multiplying by $\dfrac{1}{3}$ is the same as dividing by its reciprocal, $\dfrac{3}{1}$; and so on.

When multiplying by fractions, it can help to write whole numbers in fraction form. For example, $24 = 24 \div 1 = \dfrac{24}{1}$.

To multiply $\dfrac{2}{3}$ by 24, write 24 as $\dfrac{24}{1}$. The problem is now: $\dfrac{2}{3} \times \dfrac{24}{1}$.

Look at both denominators (3 and 1) to see if either one shares a common factor, other than 1, with either of the two numerators (2 and 24).

Here, the denominator 3 and the numerator 24 share a common factor of 3.

Divide the denominator 3 and the numerator 24 by the common factor of 3.

$$\dfrac{2}{\underset{1}{\cancel{3}}} \times \dfrac{\overset{8}{\cancel{24}}}{1}$$

Now, multiply, remembering that a fraction with a denominator of 1 can be written as a whole number.

$$\dfrac{2}{1} \times \dfrac{8}{1} = 2 \times 8 = 16$$

So, $\dfrac{2}{3} \times 24 = 16$.

MATH NOTE

When, in multiplying fractions, a common factor is removed from both a numerator and a denominator, the method is called *canceling*.

Hints for changing mixed numbers to improper fractions

A *mixed number* has a whole number part and a fraction part. The mixed number $4\frac{1}{2}$ has the whole number part 4 and the fraction part $\frac{1}{2}$. To change $4\frac{1}{2}$ to the equivalent *improper fraction*, or fraction with its numerator larger than its denominator, do the following steps:

1. Multiply the whole number by the denominator.
   $4 \times 2 = 8$

2. Add the numerator to the product.
   $1 + 8 = 9$

3. The number that results is the new numerator.
   The new numerator is **9**.

4. Use the same denominator.
   The denominator is **2**.

The improper fraction is $\frac{9}{2}$.

Hints for dividing a mixed number by a whole number

Find the quotient: $2\frac{3}{4} \div 5$.

First change $2\frac{3}{4}$ to an improper fraction. The new numerator is

$(2 \times 4) + 3 = 11$. The denominator remains 4. So, $2\frac{3}{4} = \frac{11}{4}$.

Then change 5 to a fraction with a denominator of 1: $5 = \frac{5}{1}$.
Now rewrite the problem.

$$\frac{11}{4} \div \frac{5}{1}$$

Next, change the division sign to a multiplication sign and the second fraction into its reciprocal (flip it!). Remember: Dividing by a fraction is the same as multiplying by its reciprocal.

$$\frac{11}{4} \times \frac{1}{5}$$

Now multiply the fractions! The steps are listed below.

Hints for multiplying two fractions

(Note that the steps below can be used to multiply two mixed numbers if you first write them as improper fractions.)

1. Multiply the numerators.

2. Multiply the denominators.

3. Cancel if you can.

Now you can multiply.

$$\frac{11}{4} \times \frac{1}{5} = \frac{11 \times 1}{4 \times 5} = \frac{11}{20}$$

Neither denominator of the fractions multiplied shares a common factor, other than 1, with either numerator, so $\frac{11}{20}$ is in lowest terms.

MATH NOTE

A fraction is in *lowest terms* when the only common factor shared by the numerator and denominator is one.

EXAMPLE:

In the class election, $\frac{3}{4}$ of the 24 students voted for Terry. How many students voted for Terry?

Step 2: Plan a strategy.

What operation is suggested? multiplication to find the fractional part of a whole

Estimate the answer. less than 24 but more than 12

$$\left(\frac{1}{2} \text{ of 24 is 12, and } \frac{1}{2} < \frac{3}{4} < 1 \right)$$

Step 3: Do the plan.

Multiply to find the number of students who voted for Terry.

$$\frac{3}{4} \times 24 = \frac{3}{4} \times \frac{\overset{6}{24}}{1} = \frac{18}{1} = 18$$

Eighteen students voted for Terry.

Step 4: Check your work.

Check the multiplication with division.

$$18 \div \frac{3}{4} = 18 \times \frac{4}{3} = \frac{\overset{6}{18}}{1} \times \frac{4}{3} = \frac{24}{1} = 24$$

EXAMPLE:

A sweater is on sale for $\frac{1}{3}$ off the regular price of $48. Find the sale price of the sweater.

Step 2: Plan a strategy.

What operations are suggested? multiplication to find out by how much the price is reduced; subtraction to find the sale price

Estimate the answer. more than $24 (If the sweater were $\frac{1}{2}$ off, the sale price would be $24; $\frac{1}{3}$ off is less than $\frac{1}{2}$ off.)

Alternate Strategy: Find the fraction of the regular price that is to be paid (1 regular price − $\frac{1}{3}$ off = $\frac{2}{3}$ of the regular price); multiply this fraction times the regular price.

Step 3: Do the plan.

First, multiply to find the amount off the regular price.

$$\frac{1}{3} \times 48 = \frac{1}{\cancel{3}_1} \times \frac{\cancel{48}^{16}}{1} = \frac{16}{1} = 16$$

The sweater is reduced by $16.
Subtract to find the sale price of the sweater.

$$48 - 16 = 32$$

Using the alternate strategy:

$$\frac{2}{3} \times 48 = \frac{2}{\cancel{3}_1} \times \frac{\cancel{48}^{16}}{1} = \frac{32}{1} = 32$$

The sweater is on sale for $32.

Step 4: Check your work.
Work backwards.
First check the subtraction with addition.

$$32 + 16 = 48$$

Then check the multiplication with division.

$$16 \div \frac{1}{3} = 16 \times 3 = 48$$

EXAMPLE:

Brianna has $4\frac{1}{2}$ feet of ribbon to use to wrap eight presents. If each present uses an equal amount of ribbon, how many feet of ribbon can be used to wrap each present? All of the ribbon is to be used.

Step 2: Plan a strategy.

What operation is suggested? division, since you know the total number of feet of ribbon and the number of groups (presents), and you wish to find the number of feet in one group

Estimate the answer. more than $\frac{1}{2}$ foot (Brianna has more than 4 feet of ribbon and $4\frac{1}{2} \div 8 > 4 \div 8 = \frac{1}{2}$.)

Step 3: Do the plan.

Divide to find the number of feet of ribbon available to wrap each package.

$$4\frac{1}{2} \div 8$$

Change the mixed number to an improper fraction, and write the whole number as a fraction with a denominator of 1.

$$\frac{9}{2} \div \frac{8}{1} = \frac{9}{2} \times \frac{1}{8} = \frac{9}{16}$$

(Remember, dividing by a fraction is the same as multiplying by its reciprocal.)

Because 9 and 16 have no common factor other than 1, $\frac{9}{16}$ is in lowest terms.

So $\frac{9}{16}$ of a foot of ribbon can be used for each present.

(You can change $\frac{9}{16}$ feet to inches by multiplying by 12 inches in a foot. You should get $\frac{27}{4}$ inches, or $6\frac{3}{4}$ inches.

Try it!)

Step 4: Check your work.

Check the division with multiplication.

$$\frac{9}{16} \times 8 = \frac{9}{\underset{2}{16}} \times \frac{\overset{1}{8}}{1} = \frac{9}{2}$$

Now write $\frac{9}{2}$ as a mixed number.

$$\frac{9}{2} = 9 \div 2 = 4\frac{1}{2} \ \checkmark$$

In a fraction word problem,
If to multiply is your task,
Just look at the hints
And the job will be done fast.

BRAIN TICKLERS
Set # 9

Solve each word problem by reading it carefully and thoroughly, planning a strategy, estimating the answer, doing the plan, and then
checking your work. Then in your own words,
list the steps you used to solve the problem.

1. Counselor Pearl has $12\frac{3}{4}$ feet of string to be
   shared equally by her campers. Each camper
   will receive $2\frac{1}{8}$ feet of string. How many
   campers are in Pearl's group?

2. Each of three friends was given 24 tickets to sell for the school banquet.
   Ming sold $\frac{2}{3}$ of her tickets, Matthew sold $\frac{1}{4}$ of his tickets, and Joy sold $\frac{5}{12}$
   of her tickets. How many tickets did each friend sell?

3. At the bake sale, there were a total of 64 cookies. The students sold
   $\frac{5}{8}$ of the cookies for $0.45 each and the rest for $0.60 each. How much
   money was collected if all of the cookies were sold?

4. Norton Novel bought $20\frac{5}{8}$ feet of wood to build a bookcase of five
   shelves. If each shelf requires the same amount of wood and all the
   wood is to be used, how long will each shelf be?

5. A recipe calls for $2\frac{1}{4}$ cups of flour, $\frac{3}{4}$ cup of white sugar, and $1\frac{1}{2}$ cups
   of brown sugar. If the recipe is doubled, how many cups of each will
   be needed?

(Answers are on page 102.)

We've multiplied and divided
Our fractions with success.
To add and to subtract
Is the job that is next.

Hints for adding fractions and mixed numbers

Add $15\frac{1}{2}$ and $1\frac{3}{4}$.

1. Because you cannot add halves and quarters directly, you must first find the *least common denominator* (LCD) of the fractions $\frac{1}{2}$ and $\frac{3}{4}$. This is the *least common multiple* (LCM) of the numbers 2 and 4. The least common multiple of two numbers is the smallest number into which both numbers will divide. The LCM of 2 and 4 is 4.

2. Change each fraction to an equivalent one whose denominator is the LCM, 4.

   Ask yourself, "What number do I multiply the original denominator by to get the new denominator of 4?" Multiply the numerator by this same number to get a new numerator and a resulting equivalent fraction.

   You multiply the original denominator of $\frac{1}{2}$, the 2, by 2 to get the LCM of 4. Now multiply the original numerator by 2 as well. (Remember: Multiplying both the numerator and the denominator by 2 is really multiplying by $\frac{2}{2}$, or 1.)

   $$\frac{1\times2}{2\times2} = \frac{2}{4} \qquad \frac{2}{4} \text{ is equivalent to } \frac{1}{2}.$$

   The fraction $\frac{3}{4}$ already contains the least common denominator. This fraction remains the same.

3. To add fractions with the same denominator, just add the numerators.

Do not add the denominators! The LCM is brought over as the denominator of the answer.

$$\frac{2}{4}+\frac{3}{4}=\frac{2+3}{4}=\frac{5}{4}$$

4. If the answer is an improper fraction (greater than or equal to 1), change it to a mixed number by dividing the numerator by the denominator.

$$\frac{5}{4}=5\div4=1\frac{1}{4}$$

(Notice that the remainder of 1 is expressed as a fraction with the remainder as the numerator and the denominator the same as that of the improper fraction.)

5. Add the whole number parts of the original mixed numbers: 15 + 1= 16.

The answer is $16+1\frac{1}{4}=16+1+\frac{1}{4}=17\frac{1}{4}$.

So, $15\frac{1}{2}+1\frac{3}{4}=17\frac{1}{4}$.

Hints for subtracting fractions and mixed numbers

Subtract $13\frac{5}{12}$ from $15\frac{7}{8}$.

1. If necessary, change the fractions to equivalent ones using the LCM of the denominators as the new denominator.

$$15\frac{7}{8}=15\frac{21}{24} \text{ and } 13\frac{5}{12}=13\frac{10}{24}$$

2. Subtract the numerators of the fractions to get a new numerator.

$$\frac{21}{24}-\frac{10}{24}=\frac{11}{24}$$

Do not subtract the denominators! As in addition, the LCM of the two original denominators will be the denominator of the result.

3. If possible, simplify the resulting fraction to its lowest terms by *dividing* the numerator and the denominator by their greatest common factor (GCF). (If you cannot find the GCF, divide both numbers by any common factor. Keep dividing by common factors until the only common factor of the numerator and the denominator is 1.) As 1 is the only common factor of 11 and 24, $\frac{11}{24}$ is already in lowest terms.

4. For mixed numbers, the next step is to subtract the whole numbers.

$$15 - 13 = 2$$

The answer is $2 + \frac{11}{24} = 2\frac{11}{24}$.

EXAMPLE:

On Monday, one share of Kooky Kola stock had a value of $15\frac{1}{2}$ points. On Tuesday, the stock increased $1\frac{3}{4}$ points. Find the value on Tuesday of one share of Kooky Kola stock.

Step 2: Plan a strategy.

What operation is suggested? addition, since the stock increased on Tuesday

Estimate the answer. more than 17 points, $\left(\frac{3}{4} + \frac{1}{2} \text{ is} \right.$ greater than 1, and so $\frac{1}{2} + 1\frac{3}{4}$ is greater than 2. $\left. \right)$

Step 3: Do the plan.

You can do the addition using a vertical format.

$$15\frac{1}{2} = 15\frac{2}{4}$$
$$+ \ 1\frac{3}{4} = \ 1\frac{3}{4}$$
$$\overline{} \quad \overline{}$$
$$16\frac{5}{4} = 16 + \frac{4}{4} + \frac{1}{4} = 16 + 1 + \frac{1}{4} = 17\frac{1}{4}$$

The value of one share of Kooky Kola stock on Tuesday is $17\frac{1}{4}$ points.

Step 4: Check your work.
Check your addition with subtraction.

$$17\frac{1}{4} = 16\frac{5}{4}$$
$$-1\frac{3}{4} = -1\frac{3}{4}$$
$$\overline{}$$
$$15\frac{2}{4} = 15\frac{1}{2} \quad \checkmark$$

MATH NOTE

In the immediately preceding subtraction, the fractional parts had the same denominator, but the numerator of the first fraction was less than the numerator of the second fraction. When this is the case, before you can subtract, the first mixed number must be changed to an equivalent one with a larger numerator.

$$17\frac{1}{4} = 17 + \frac{1}{4} = 16 + 1 + \frac{1}{4} = 16 + \frac{4}{4} + \frac{1}{4} = 16\frac{5}{4}$$

Thus, $17\frac{1}{4}$ was written as the equivalent $16\frac{5}{4}$.

EXAMPLE:

Carlos jogged $15\frac{7}{8}$ miles last week

and $13\frac{5}{12}$ miles this week.

How many more miles did he jog

last week than this week?

Step 2: Plan a strategy.
What operation is suggested? subtraction, as indicated by the phrase *how many more*
Estimate the answer. about 2 miles more

Step 3: Do the plan.

$$15\frac{7}{8} - 13\frac{5}{12}$$

Write the subtraction vertically.

The least common denominator of 8 and 12 is 24. Write each fraction as an equivalent fraction with a denominator of 24.

$$15\frac{7}{8} = \quad 15\frac{21}{24}$$
$$-13\frac{5}{12} = -13\frac{10}{24}$$
$$\overline{\qquad} \quad \overline{2\frac{11}{24}}$$

Step 4: Check your work.

Check your subtraction with addition. The LCM of 12 and 24 is 24.

$$2\frac{11}{24} = \quad 2\frac{11}{24}$$
$$+13\frac{5}{12} = -13\frac{10}{24}$$
$$\overline{\qquad} \quad \overline{15\frac{21}{24}}$$

The GCF of 21 and 24 is 3. Reduce the fraction $\frac{21}{24}$ to lowest terms.

$$15\frac{21}{24} = 15\frac{7}{8} \quad \checkmark$$

BRAIN TICKLERS
Set # 10

Solve each word problem by planning and carrying out a strategy. Remember to check your work. At the end of each problem you will find a letter. By placing this letter on the line above the corresponding answer in the answer code at the end of this problem set, you will be able to answer this question:

What famous cookie did Ruth Wakefield create?

FRACTION OPERATIONS

1. An ant was climbing up a 24-foot tall tree. The first hour, it climbed $\frac{1}{6}$ of the way up, the second hour, it climbed $\frac{1}{4}$ of the rest of the way up; and the third hour, it climbed $\frac{1}{5}$ of the remainder of the way up. What fractional part of the tree's height was left to climb? (O)

2. On Monday, Luis bought 50 shares of Circuit stock for $\$15\frac{1}{4}$ per share. During the week, the price of one share rose $\$1\frac{1}{8}$, went down $\$\frac{3}{16}$, and up again $\$1\frac{3}{8}$. At the end of the week, what was the value of Luis' shares of Circuit stock? (S)

3. During one day of touring in Paris, Fran walked $3\frac{2}{3}$ miles to the Louvre, $2\frac{4}{5}$ miles to the Musee D'Orsay, and $1\frac{5}{6}$ miles to the Eiffel Tower. She took a taxi to other places. How many miles did Fran walk during her day of touring? (L)

4. In training for a marathon, Danika ran $11\frac{1}{4}$ miles on Tuesday and $9\frac{5}{8}$ miles on Thursday. How many more miles did she run on Tuesday than on Thursday? (T)

5. How many $\frac{3}{4}$-inch-long beads are needed to make a 12-inch-long necklace? (H)

6. A recipe, yielding three dozen cookies, requires $\frac{1}{3}$ cup of butter, $2\frac{1}{2}$ cups of flour, and $\frac{3}{4}$ cup of sugar. If the recipe were increased to yield 48 cookies, how many combined cups of the three ingredients would be needed? (O)

7. For a barbecue, Bette Berger bought $5\frac{5}{8}$ pounds of hamburger and $6\frac{1}{4}$ pounds of hot dogs. How many more pounds of hot dogs than hamburgers did she buy? (E)

8. Cameron budgets his $840 after-tax weekly salary as follows:

   $\frac{1}{10}$ to his savings account, $\frac{2}{7}$ for his share of the rent, $\frac{1}{5}$ for school loan repayment, $\frac{1}{6}$ for utilities, and the rest for other expenses. How much of Cameron's weekly salary is left for other expenses? (L)

9. Rhoda wants to hang three pictures on a $12\frac{1}{3}$-foot-long wall. One picture is $2\frac{3}{4}$ feet long, another is $3\frac{1}{2}$ feet long, and the other is $4\frac{7}{12}$ feet long. If she hangs all three pictures on the wall, what is the combined length of the uncovered wall space? (U)

10. One foot is about $\frac{3}{10}$ of a meter. About how many meters long is a football field, which measures 100 yards long? (H)

11. A photograph measuring $3\frac{1}{2}$ inches by 5 inches is increased to a size of $8\frac{3}{4}$ inches by $12\frac{1}{2}$ inches. By how many times was the size of each dimension of the picture increased? (T)

12. What fractional part of one day is one minute? (E)

Answer Code

$$\overline{\quad} \quad \overline{\quad} \quad \overline{\quad} \qquad \overline{\quad} \quad \overline{\quad} \quad \overline{\quad} \quad \overline{\quad} \qquad \overline{\quad} \quad \overline{\quad} \quad \overline{\quad}$$

$1\frac{5}{8}$ 16 $\frac{5}{8}$ $2\frac{1}{2}$ $\frac{1}{2}$ 208 $12\frac{3}{10}$ 90 $4\frac{7}{9}$ $1\frac{1}{2}$

$$\overline{\qquad\qquad} \quad \overline{\qquad\qquad} \text{ cookie.}$$

$878\frac{1}{8}$ $\frac{1}{1440}$

(Answers are on page 104.)

BRAIN TICKLERS—THE ANSWERS

Set # 7, page 74

1. Estimate: about $300 (about $8 per hour × 7 hours per day = $56 per weekday; about $56 per weekday × 5 weekdays = about $300 for 5 weekdays; about $9 per hour × 7 hours = $63 for one Saturday; and $300 + $63 = $363)
First, multiply to find how much Victor earns each weekday.

$$7 \times \$7.95 = \$55.65$$

Then multiply to find how much he earns in 5 weekdays.

$$5 \times \$55.65 = \$278.25$$

Now multiply to find how much he earns for working on Saturday.

$$7 \times 9.25 = \$64.75$$

Now add to find how much Victor earns in total for working 5 weekdays and one weekend day.

$$\begin{array}{r} \$278.25 \\ + \$\ 64.75 \\ \hline \$343.00 \end{array}$$

Victor earns $343.00 for working 5 weekdays and one weekend day.

Check your answer by working backwards.

Check your addition by subtraction.

$$
\begin{array}{r}
343.00 \\
- \ 64.75 \\
\hline
278.25
\end{array}
\qquad \text{or} \qquad
\begin{array}{r}
343.00 \\
- \ 278.25 \\
\hline
64.75 \quad \checkmark
\end{array}
$$

Now check your last multiplication by division.

$$
\begin{array}{r}
9.25 \ \checkmark \\
7,00.\overline{)64,75.00} \\
6300 \\
\hline
1750 \\
1400 \\
\hline
3500 \\
3500 \\
\hline
0
\end{array}
$$

Now check your first two multiplications.

$$
\begin{array}{r}
55.65 \ \checkmark \\
5,00.\overline{)278,25.00} \\
2500 \\
\hline
2825 \\
2500 \\
\hline
3250 \\
3000 \\
\hline
2500 \\
2500 \\
\hline
0
\end{array}
\qquad
\begin{array}{r}
7.95 \ \checkmark \\
7,00.\overline{)55,65.00} \\
4900 \\
\hline
6650 \\
6300 \\
\hline
3500 \\
3500 \\
\hline
0
\end{array}
$$

CROSS OUT 343 IN THE SOLUTION BOX.

2. Estimate: more than 18 (36 ÷ 2 = 18; 1.75 < 2)
    First convert one yard to inches.

$$1 \text{ yd} = 36 \text{ in.}$$

Then divide to find how many groups of 1.75 inches are in a total of 36 inches.

36 ÷ 1.75 ≈ 20.57 (The symbol "≈" means "approximately equal to.")

Now round to the nearest whole number.
21 Oreos laid end to end would form a line about one yard long.

Check: Work backwards.
Check division with multiplication.

$$20.57 \text{ Oreos} \times 1.75 \text{ inches per Oreo} \approx 36.0 \quad \checkmark$$

CROSS OUT 21 IN THE SOLUTION BOX.

3. Estimate: about $24 (300 words = 30 groups of 10; $30 \times 0.25 \approx$
$8; 8 sheets $\times$ \$2.25 $\approx$ \$16; \$8 + \$16 = \$24)
First, find the cost to print 300 words.

$$300 \div 10 = 30 \text{ groups of } 10$$

$$\$0.25 \times 30 = \$7.50$$

Next, find the cost of the 8 sheets of paper.

$$\$2.25 \times 8 = \$18.00$$

Now add to find the total charge.

$$\$7.50 + \$18.00 = \$25.50$$

The calligrapher would charge $25.50.

Check your work by working backwards.
First check addition with subtraction.

$$\$25.50 - \$18.00 = \$7.50 \quad \text{or} \quad \$25.50 - \$7.50 = \$18.00 \quad \checkmark$$

Then check each multiplication with division.

$$\$18.00 \div 8 = \$2.25 \quad \text{or} \quad \$18.00 \div \$2.25 = \$8 \quad \checkmark$$

$$\$7.50 \div \$0.25 = 30 \quad \text{or} \quad \$7.50 \div 30 = \$0.25 \quad \checkmark$$

Last, check the first division by multiplication.

$$30 \times 10 = 300 \quad \checkmark$$

CROSS OUT 25.50 IN THE SOLUTION BOX.

4. Estimate: between 26 and 52 kilometers ($1 \leq 1.6 \leq 2$; $1 \times 26 = 26$; $2 \times 26 = 52$)
First multiply the number of miles by 1.6 to determine the number of kilometers.

$$
\begin{array}{r}
26.2 \\
\times\, 1.6 \\
\hline
1572 \\
2620 \\
\hline
41.92
\end{array}
\quad (1 + 1 = \text{spaces back})
$$

The number of kilometers in 26.2 miles is about 41.92.

Check your work.
Check multiplication with division.

$$41.92 \div 1.6 = 26.2 \quad \text{or} \quad 41.92 \div 26.2 = 1.6 \checkmark$$

CROSS OUT 41.92 IN THE SOLUTION BOX.

5. Estimate: about 175 centimeters ($180 - 20 = 160$; $160 + 15 = 175$; $175 + 0 = 175$)
Rachel is 17.6 centimeters shorter than Joshua. Therefore, subtract to find Rachel's height.

$$181.0 \text{ cm} - 17.6 \text{ cm} = 163.4 \text{ cm}$$

Corbin is 15.2 centimeters taller than Rachel. Add to find Corbin's height.

$$163.4 + 15.2 \text{ cm} = 178.6 \text{ cm}$$

Janet is taller than Corbin. Add to find Janet's height.

$$178.6 \text{ cm} + 0.4 \text{ cm} = 179.0 \text{ cm}$$

Janet is 179.0 centimeters tall.

Check your work by working backwards.
Check the last addition with subtraction to find Corbin's height.

$$179.0 \text{ cm} - 0.4 \text{ cm} = 178.6 \text{ cm}$$

Next, check the previous addition with subtraction to find Rachel's height.

$$178.6 \text{ cm} - 15.2 \text{ cm} = 163.4 \text{ cm}$$

Now check the subtraction with addition to find Joshua's height.

$$163.4 \text{ cm} + 17.6 \text{ cm} = 181.0 \text{ cm} \checkmark$$

CROSS OUT 179.0 IN THE SOLUTION BOX.

6. Estimate: about $65 (220 − 170 = 50; 50 + 35 = 85; 85 − 20 = 65)
   Subtract the first withdrawal from the initial savings.

$$
\begin{array}{r}
218.35 \\
- \ 174.50 \\
\hline
43.85
\end{array}
$$

Next, add the amount of the check deposited.

$$
\begin{array}{r}
43.85 \\
- \ 34.99 \\
\hline
78.84
\end{array}
$$

Last, subtract the last withdrawal.

$$
\begin{array}{r}
78.84 \\
- \ 18.25 \\
\hline
60.59
\end{array}
$$

There was a balance of $60.59 remaining in Samuel's account.

Check by working backwards.
Check the last subtraction with addition.

$$60.59 + 18.75 = 78.84$$

Next, check the addition with subtraction.

$$78.84 - 34.99 = 43.85 \quad \text{or} \quad 78.84 - 43.85 = 34.99$$

Last, check the first subtraction with addition.

$43.85 + 174.50 = 218.35$ (the initial amount in Samuel's account) $\checkmark$

CROSS OUT 60.59 IN THE SOLUTION BOX.

7. Estimate: more than 3 minutes but less than 11 minutes
(3 minute call costs $0.35; 11 minute call costs
$8 \times 0.25 + 0.35 = \$2.35; 2.35 > 1.85$)
First, subtract $0.35 for the first three minutes.

$$\$1.85 - \$0.35 = \$1.50$$

Then, divide the remaining charge by the charge per minute
for over three minutes.

$$\$1.50 \div \$0.25 = 6 \text{ min}$$

Last, add to find the total time.

$$3 \text{ min} + 6 \text{ min} = 9 \text{ min}$$

The call was 9 minutes long.

Check your work by working backwards.
First check addition with subtraction.

$$9 - 6 = 3$$

Then check the division by multiplication.

$$6 \times \$0.25 = \$1.50$$

Last, check subtraction with addition.

$$\$1.50 + \$0.35 = \$1.85 \text{ (total charge of the call)} \checkmark$$

CROSS OUT 9 IN THE SOLUTION BOX.

8. Estimate: hard to estimate, but cannot be more than 9, since
9 candy bars were bought altogether
This problem can be solved by the guess and check method.
Keep a careful list of your guesses and checks until the conditions of the problem have been met. That is, there must be a
total of 9 candy bars and a total cost of $4.45.

Number of Crunchy bars	Total for Crunchy bars	Number of Wonder bars	Total for Wonder bars	Total Cost	
3	$3 \times \$0.45 = \$1.35$	6	$6 \times \$0.55 = \$3.30$	$4.65	
4	$4 \times \$0.45 = \$1.80$	5	$5 \times \$0.55 = \$2.75$	$4.55	
5	$5 \times \$0.45 = \$2.25$	4	$4 \times \$0.55 = \$2.20$	$4.45	✓

Candace bought five Crunchy bars.

Check your multiplications and additions to make sure your calculations are correct.

CROSS OUT 5 IN THE SOLUTION BOX.

9. Estimate: about 13 inches (3 years × 365 days is about 1,000 days; $1,000 \times 0.013$ inches = 13 inches)
First, multiply to find the total number of days in three non-leap years.

$$3 \text{ years} \times 365 \text{ days per year} = 1,095 \text{ days}$$

Next, multiply the number of days by the growth per day.

$$1,095 \text{ days} \times 0.013 \text{ inches per day} = 14.235$$

Round to the nearest hundredth: 14.24.
Your hair would grow about 14.24 inches.

Check your work by working backwards.
First check the last multiplication with division.

$$14.24 \div 1,095 = 0.013 \ ✓$$

Then check the first multiplication with division.

$$1,095 \div 3 = 365 \text{ (the number of days in a year)} \ ✓$$

CROSS OUT 14.24 IN THE SOLUTION BOX.

10. Estimate: about one more gallon (7 liters is about $7 \times 0.3 \approx 2$ gallons; 3 gallons – 2 gallons = 1 gallon) First, multiply to find the number of gallons of juice Reggie bought.

$$7 \text{ liters} \times 0.26 \text{ gallons per liter} = 1.82 \text{ gallons}$$

Then, subtract to find how many more gallons Sarah brought than Reggie.

$$3 \text{ gallons} - 1.82 \text{ gallons} = 1.18 \text{ gallons}$$

Sarah brought 1.18 more gallons of juice than Reggie brought. Reggie was not correct.

Check your work by working backwards.
First check subtraction with addition.

$$1.18 + 1.82 = 3.00 \checkmark$$

Then check multiplication with division.

$$1.82 \div 0.26 = 7 \quad \text{or} \quad 1.82 \div 7 = 0.26 \checkmark$$

CROSS OUT 1.18 IN THE SOLUTION BOX.

The only number left in the Solution Box is 22.59. The heaviest ice cream sundae ever made weighed 22.59 tons!

Set # 8, page 76

1. a. Add the costs of Paul's purchases: $15.95 + $15.95 + $7.89 + $49.75 = $89.54. Looking at the shipping chart, $89.74 is between $76.00 and $99.99, so the shipping charge will be $11.95. Add the shipping charges to the purchase price: $89.54 + $11.95 = $101.49. Paul's gift of $100 will not be enough to pay for his purchase. How much more is needed? Subtract to find out: $101.49 - $100.00 = $1.49 more.
    b. Guess and check for this one! Gila must have spent under $100, as a purchase of $100 or more would result in a shipping charge of $15, which would make the final cost higher than her cost of $111.51. Her purchase could not have been $75.99 or less, as the shipping charge of, at the most, $10.95 would only add up to $75.99 + $10.95 = $86.94, less than her cost. Thus, her purchases were in the range of

$76.00 to $99.99. Subtract the shipping charge of $11.95 from her total of $111.51: $111.51 − $11.95 = $99.56 for the four items. Add up pairs of different items and then double the cost since Gila bought two each of two items. There are only two items that doubled give a total cost of $99.56. Gila bought two pasta sets at $35.49 × 2 = $70.98 and two disk caddies at $14.29 × 2 = $28.58. Now $70.98 + $28.58 = $99.56, and $99.56 + $11.95 (shipping) = $111.51, which is correct.

2. a. The check payment is subtracted from the account. To find the check payment, subtract the balance of $293.59 from the previous balance of $329.42 to get $35.83.
    b. The deposit is added to the account. To find the new balance, add the deposit of $145.85 to the previous balance of $293.59 to get $439.44.
    c. The withdrawal is subtracted from the account. To find the withdrawal amount, subtract the final balance of $373.94 from the previous balance of $439.44 to get $65.50.

3. a. First look at the whole number parts of the decimals. Because 22 is the largest, 22.47 (Timmerman) has the greatest distance. The other distances are between 21 and 22 meters. Find the decimal with the largest value in the tenths' place (21.70), as this is the next greatest distance. Order the remaining decimals from the one with the largest value in the tenths place to the smallest value. The full list, in order, follows: 22.47 (Timmerman), 21.70 (Tulce), 21.62 (Barnes), 21.35 (Kyselov), and 21.26 (Andrei).
    b. The word "difference" means to subtract. Subtract the least distance from the greatest distance. The difference between the greatest shot put distance and the least distance is 22.47 − 21.26 = 1.21 meters.
    c. Look at the chart to find consecutive Olympic years in which there was an increase in distance. From 1980 to 1984 there was a decrease in distance. The increase in distance from 1984 to 1988 was 22.47 − 21.26 = 1.21 meters. The other sets of consecutive Olympic years all had a decrease in distance; thus, 1984 to 1988 had the greatest and only increase in distance.

Set # 9, page 85

1. Divide the total length of string by the length to be shared equally by each camper.

$$12\frac{3}{4} \div 2\frac{1}{8}$$

Change each mixed number to its equivalent improper fraction. Thus, $12\frac{3}{4} = 5\frac{1}{4}$ and $2\frac{1}{8} = 1\frac{7}{8}$.

$$12\frac{3}{4} \div 2\frac{1}{8} = 5\frac{1}{4} \div 1\frac{7}{8}$$

Change the division to multiplication by using the reciprocal of $1\frac{7}{8}$, which is $\frac{8}{17}$.

$$\frac{51}{4} \div \frac{17}{8} = \frac{\overset{3}{\cancel{51}}}{\underset{1}{\cancel{4}}} \times \frac{\overset{2}{\cancel{8}}}{\underset{1}{\cancel{17}}} = \frac{6}{1} = 6$$

There are 6 campers in Pearl's group.

2. Multiply the fractional portion of the tickets sold by each friend by the number of tickets to be sold.
   For Ming,

$$\frac{2}{3} \times 24 = \frac{2}{\underset{1}{\cancel{3}}} \times \frac{\overset{8}{\cancel{24}}}{1} = \frac{16}{1} = 16 \text{ tickets}$$

For Matthew,

$$\frac{1}{4} \times 24 = \frac{1}{\underset{1}{\cancel{4}}} \times \frac{\overset{6}{\cancel{24}}}{1} = \frac{6}{1} = 6 \text{ tickets}$$

For Joy,

$$\frac{5}{12} \times 24 = \frac{5}{\underset{1}{\cancel{12}}} \times \frac{\overset{2}{\cancel{24}}}{1} = \frac{10}{1} = 10 \text{ tickets}$$

3. First find the number of cookies sold at $0.45 each.

$$\frac{5}{8} \times 64 = \frac{5}{\underset{1}{\cancel{8}}} \times \frac{\overset{8}{\cancel{64}}}{1} = \frac{40}{1} = 40 \text{ cookies}$$

Find the amount of money collected from the sale of $0.45 cookies.

$$40 \text{ cookies} \times \$0.45 \text{ per cookie} = \$18.00$$

Find the number of cookies sold at $0.60 each. Since 64 cookies were sold altogether and 40 cookies were sold at $0.45 each, the difference 64 − 40 = 24 is the number of cookies sold at $0.60 each.
Now find the amount of money collected from the sale of $0.60 cookies.

$$24 \text{ cookies} \times \$0.60 \text{ per cookie} = \$14.40$$

The total amount of money collected from the sale of the cookies is $18.00 + $14.40 = $32.40.

4. Drawing a picture may help you discover the operation needed.

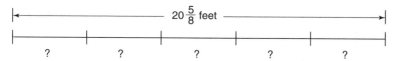

Divide the total length by 5 to find the length of one shelf. Remember, you divide when you know the total (length of the board) and the number of groups (shelves), and wish to find the number in each group (length of a shelf).

$$20\frac{5}{8} \div 5 = \frac{165}{8} \div 5 = \frac{\overset{33}{\cancel{165}}}{8} \times \frac{1}{\cancel{5}_1} = \frac{33}{8} = 4\frac{1}{8}$$

(Remember that the reciprocal of 5 is $\frac{1}{5}$, and that dividing is the same as multiplying by the reciprocal.)

The length of one shelf is $4\frac{1}{8}$ feet.

5. To double a recipe means to multiply all amounts of ingredients by two. Find the amount of flour.

$$2\frac{1}{4} \times 2 = \frac{9}{\underset{2}{\cancel{4}}} \times \frac{\cancel{2}^1}{1} = \frac{9}{2}$$

Find the amount of white sugar.

$$\underset{2}{\overset{1}{\cancel{\frac{3}{4}}}} \times \frac{\overset{1}{\cancel{2}}}{1} = \frac{3}{2} = 1\frac{1}{2} \text{ cups}$$

Find the amount of brown sugar. Change $1\frac{1}{2}$ to $\frac{3}{2}$ before doubling.

$$1\frac{1}{2} \times 2 = \underset{1}{\overset{3}{\cancel{\frac{3}{2}}}} \times \frac{\overset{1}{\cancel{2}}}{1} = \frac{3}{1} = 3 \text{ cups}$$

Set # 10, page 90

1. This problem involves many steps. You need to find the number of feet climbed by the ant each of the three hours. In the first hour, the ant climbed $\frac{1}{6}$ of the height of the tree. Thus, in the first hour the ant climbed $\frac{1}{6} \times 24$ feet = 24 feet $\div$ 6 = 4 feet. There are now 24 feet – 4 feet = 20 feet left to climb. In the second hour the ant climbed $\frac{1}{4}$ of the remaining way up the tree. Thus, in the second hour the ant climbed an additional $\frac{1}{4} \times 20$ feet = 20 feet $\div$ 4 = 5 feet. There are now 20 feet – 5 feet = 15 feet left to climb. In the third hour, the ant climbed $\frac{1}{5}$ of the remaining way up the tree. Thus, in the third hour the ant climbed $\frac{1}{5} \times 15$ feet = 15 feet $\div$ 5 = 3 feet. There are now 15 feet – 3 feet = 12 feet left to climb. Now 12 feet out of a total of 24 feet is the fraction $\frac{12}{24}$, which simplifies to $\frac{1}{2}$. The ant has $\frac{1}{2}$ of the tree's height left to climb.

   Put an O on the line above $\frac{1}{2}$.

2. To find the value of one share of stock, add the increases and subtract the decreases. The stock started at $15\frac{1}{4}$, rose by $1\frac{1}{8}$, went down by $\$\frac{3}{16}$, and then rose again by $1\frac{3}{8}$.

The value of one share of the stock was then $15\frac{1}{4} + 1\frac{1}{8} - \frac{3}{16} + 1\frac{3}{8}$. The least common denominator of the fractions is 16.

$$15\frac{4}{16} + 1\frac{2}{16} - \frac{3}{16} + 1\frac{6}{16} = 17\frac{9}{16}$$

To get the value of Luis' stock, multiply the value of one share of stock by the number of shares.

$$17\frac{9}{16} \times 50 = \frac{281}{8} \times \frac{\overset{25}{\cancel{50}}}{1} = \frac{7,025}{8} = 878\frac{1}{8}$$

The value of Luis' shares of Circuit stock at the end of the week was $\$878\frac{1}{8}$.

Put an S on the line above $\$878\frac{1}{8}$.

3. Because you are finding the total number of miles Fran walked, you need to add up all of the fractions. First change the fractions to fractions with a common denominator of 30, the LCM of 3, 5, and 6.

$$3\frac{20}{30} + 2\frac{24}{30} + 5\frac{25}{30} = 10\frac{69}{30}$$

Convert $\frac{69}{30}$ to its mixed number form of $2\frac{9}{30}$, which simplifies to $2\frac{3}{10}$.

The final answer is $10 + 2 + \frac{3}{10} = 12\frac{3}{10}$.

Fran walked $12\frac{3}{10}$ miles.

Put an L on the line above $12\frac{3}{10}$.

4. To find how many more miles Danika ran on Tuesday than on Thursday, subtract the number of miles she ran on Thursday from the number of miles she ran on Tuesday.

$$11\frac{1}{4} - 9\frac{5}{8} = 11\frac{2}{8} - 9\frac{5}{8}$$

Because $\frac{5}{8} > \frac{2}{8}$, subtract 1 from the 11 in $11\frac{2}{8}$ and add it back as $\frac{8}{8}$ (1) to the fraction part $\frac{2}{8}$. Thus, $11\frac{2}{8}$ is equivalent to $10\frac{10}{8}$.

$$10\frac{10}{8} - 9\frac{5}{8} = 1\frac{5}{8}$$

Danika ran $1\frac{5}{8}$ miles more on Tuesday than she did on Thursday.

Put a T on the line above $1\frac{5}{8}$.

5. Drawing a picture may help you discover the operation needed.

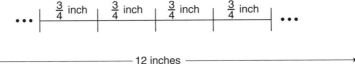

Divide, since you know the total number of inches (length of the necklace) and the number of inches in each group (bead), and wish to find the number of groups (beads).

$$\frac{12}{1} \div \frac{3}{4} = \frac{\overset{4}{\cancel{12}}}{1} \times \frac{4}{\underset{1}{\cancel{3}}} = \frac{16}{1} = 16$$

There are 16 beads in the necklace.
Put an H on the line above 16.

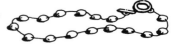

6. The given recipe makes three dozen cookies. Three dozen times 12 cookies per dozen equals 36 cookies. To make 48 cookies, each ingredient must be increased. When you increase the ingredients in a recipe you multiply the amount of each ingredient by a common number to increase every ingredient at the same rate. You need to find how many times larger 48 is than 36. Divide to find out: $48 \div 36 = 1\frac{1}{3}$. Each

amount must be multiplied by $1\frac{1}{3}$ $\left(\frac{4}{3}\right)$. You will need

$\frac{1}{3} \times \frac{4}{3} = \frac{4}{9}$ cup of butter; $\frac{5}{2} \times \frac{\overset{2}{\cancel{4}}}{3} = \frac{10}{3} = 3\frac{1}{3}$ cups of flour;

$1\frac{3}{4} \times 4\frac{1}{3} = 1$ cup of sugar. Now add all of the amounts to find the
total number of cups of all ingredients needed to make 48 cookies.

$$\frac{4}{9} + 3\frac{1}{3} + 1 = \frac{4}{9} + \frac{10}{3} + 1 = \frac{4}{9} + \frac{30}{9} + \frac{9}{9} = \frac{43}{9} = 4\frac{7}{9} \text{ cups}$$

Put an O on the line above $4\frac{7}{9}$.

7. To find how many more pounds of hot dogs than pounds of hamburger Bette bought, subtract the number of pounds of hamburger from the number of pounds of hot dogs.

$$6\frac{1}{4} - 5\frac{5}{8} = 6\frac{2}{8} - 5\frac{5}{8} = 5\frac{10}{8} - 5\frac{5}{8} = \frac{5}{8}$$

Bette bought $\frac{5}{8}$ pound more of hot dogs than hamburger.

Put an E on the line above $\frac{5}{8}$.

8. First find the amount Cameron budgets to his savings account, rent, school loan repayment, and utilities.

For savings: $\frac{1}{10} \times \$840 = \$840 \div 10 = \$84$

For rent: $\frac{2}{7} \times \$840 = \frac{2}{1\!\!\!/7} \times \frac{\overset{120}{\cancel{\$840}}}{1} = \$240$

For school loan repayment: $\frac{1}{5} \times \$840 = \$840 \div 5 = \$168$

For utilities: $\frac{1}{6} \times \$840 = \$840 \div 6 = \$140$

Add the amounts that Cameron budgets for these expenses.

$$\$84 + \$240 + \$168 + \$140 = \$632$$

This leaves $\$840 - \$632 = \$208$ for other expenses. Cameron has \$208 left from his weekly salary for other expenses.
Put an L on the line above 208.

9. In order to find the length of uncovered space, first find the total length of the covered space by adding the lengths of all three pictures. Rewriting each fraction with a common denominator of 12, the sum is

$$2\frac{9}{12}+3\frac{6}{12}+4\frac{7}{12}=9\frac{22}{12}=10\frac{10}{12}, \text{ or } 10\frac{5}{6}.$$

Subtract this amount from the length of the wall.

$$12\frac{2}{6}-10\frac{5}{6}=11\frac{8}{6}-10\frac{5}{6}=1\frac{3}{6}=1\frac{1}{2}$$

There is a combined length of $1\frac{1}{2}$ feet of uncovered wall space.

Put a U on the line above $1\frac{1}{2}$.

10. 100 yards × 3 feet per yard = 300 feet;

    300 feet × $\frac{3}{10}$ meter per foot = 90 meters.

    A football field 100 yards long is about 90 meters long.
    Put an H on the line above 90.

11. The length and width of the picture are each increased by the same factor. To find the factor of increase (the same method as in increasing recipes), you need only divide one set of numbers. (However, you may divide the other numbers for a check.) The problem is

    $$8\frac{3}{4}\div3\frac{1}{2}=\frac{35}{4}\div\frac{7}{2}=\frac{535}{4}\times\frac{21}{7}=\frac{5}{2}=2\frac{1}{2}$$

    Each dimension was increased $2\frac{1}{2}$ times.

    Put a T on the line above $2\frac{1}{2}$.

12. 60 minutes per hour × 24 hours per day is 1,440 minutes per day. One minute is $\frac{1}{1,440}$ of one day.

    Put an E on the line above $\frac{1}{1,440}$.

Answer Code

T	H	E	T	O	L	L	H	O	U	S	E
$1\frac{5}{8}$	16	$\frac{5}{8}$	$2\frac{1}{2}$	$\frac{1}{2}$	208	$12\frac{3}{10}$	90	$4\frac{7}{9}$	$1\frac{1}{2}$	$878\frac{1}{8}$	$\frac{1}{1,440}$

Ruth Wakefield created the Toll House cookie.

CHAPTER FOUR

Rewarding Ratios and Rates, and Prize-Winning Proportions

Comparing two quantities
As a ratio, proportion, or rate,
Makes word problems a snap!
So let's go, don't wait!

WORD PROBLEMS WITH RATIOS

When you compare two quantities with the same units, you are finding a *ratio*. To solve word problems with ratios, you will again use Polya's four steps. Remember, before you can plan a strategy, you must understand the problem. To understand a problem with ratios, you must first read the problem carefully. Then ask yourself these questions: What is the problem about? What are you being asked to find? Is there enough information?

You must also plan a strategy, do your plan, and check your answer.

Hints for finding ratios

Before solving a problem with ratios, here are some hints for finding a ratio.

To find the ratio of 30 miles to 45 miles, use the following steps. First look to make sure that the units are the same. In this case, the units are miles and miles.

Now look for the corresponding numbers. In this case, the numbers 30 and 45.

Create a fraction by making the number before the word "to" in the comparison the numerator and the number after the word "to" the denominator. Here, the fraction is $\frac{30 \text{ mi}}{45 \text{ mi}}$.

Express the fraction in simplest form. (Refer to Chapter 3 for hints on equivalent fractions and simplest form.)

$$\frac{30}{45} = \frac{2}{3} = 2 \text{ to } 3$$

The ratio $\frac{30}{45}$ is the same as $\frac{2}{3}$, which is read as the ratio "two to three," and not as the fraction "two-thirds."

MATH NOTE

In a problem involving ratios, the fraction bar is read as "to." The ratio is not read as you would read a traditional fraction.

Ratios may be written:
    as a fraction in simplest form, or
    as a phrase using the word "to" between the numbers, or

with a colon between the numbers (as in 2 : 3). The colon is read as the word "to."

In this book we will often express ratios as fractions.

EXAMPLE:

In training for the marathon, Max ran 30 miles in week one and 45 miles in week two. What is Max's ratio of miles from week one to week two?

Step 2: Plan a strategy.

Find the ratio of distance run in week one to week two; use your understanding of fractions to form and reduce the ratio.

Step 3: Do the plan.

First check that the units used are the same for each distance: miles and miles.

Be careful to put the numbers in the correct part of the ratio. In this case, the ratio is $\frac{30 \text{ mi}}{45 \text{ mi}}$, which reduces to $\frac{2}{3}$ (2 to 3).

Step 4: Check your work.

In this case, checking means going over the problem again. Make sure you have the numbers in the right order and that the fraction is correctly reduced. Notice that if you had written $\frac{45}{30}$ and simplified to $\frac{3}{2}$ or 3 to 2, you would not be answering the question correctly. You would be comparing week two to week one and not, as asked, week one to week two.

EXAMPLE:

Last season the Planets basketball team won 15 games and lost 25 games. Find the ratio of games won to total games played.

Step 2: Plan a strategy.
First use addition to find the total number of games played last season.
Find the ratio of games won to total games played; use your understanding of fractions to form and reduce the ratio.

Step 3: Do the plan.
Add the wins and losses to find the total number of games played: 15 + 25 = 40 games played.
Check that the units used are the same for each number: games and games. ✓
Be careful to put the numbers in the correct part of the ratio. In this case, which asks for the ratio of the number of games won (15) to the total number of games played (40), the ratio is $\frac{15 \text{ games}}{40 \text{ games}}$, which reduces to $\frac{3}{8}$ (3 to 8).

Step 4: Check your work.
In this case, checking means going over the problem again. Make sure that you have the correct total number of games, that the numbers are in the right order, and that the fraction is correctly reduced.

EXAMPLE:

The movie in Cinema 1 is 90 minutes long and the movie in Cinema 2 is 2 hours long. Find the ratio of the length of the movie in Cinema 1 to the length of the movie in Cinema 2.

Step 2: Plan a strategy.
First convert the length in hours of the movie in Cinema 2 to minutes.
Then find the ratio of the length in minutes of the movie in Cinema 1 to the length in minutes of the movie in Cinema 2; use your understanding of fractions to form and reduce the ratio.

Step 3: Do the plan.

The length in minutes of the movie in Cinema 2 is 2 hours × 60 minutes per hour = 120 minutes.

Check that the units used are the same for each number: now minutes to minutes. ✓

Be careful to put the numbers in the correct part of the ratio. In this case, which asks for the ratio of the length of the movie in Cinema 1 to the length of the movie in Cinema 2, the ratio is $\frac{90 \text{ min}}{120 \text{ min}}$, which reduces to $\frac{3}{4}$ (3 to 4).

Step 4: Check your work.

In this case, checking means going over the problem again. Make sure that you have changed hours to minutes correctly, that the numbers are in the right order, and that the fraction is correctly reduced.

MATH NOTE

When a ratio involves a common measurement type (distance, time, money, weight, and so on) but the units are not the same, change one of the units to match the other one. It is usually easier to change the larger unit to the smaller unit.

For example, to change a ratio of *hours to minutes* to *minutes to minutes*, multiply the number of hours by 60.

To change *dollars to cents* to *cents to cents*, multiply the number of dollars by 100.

To change *feet to inches* to *inches to inches*, multiply the number of feet by 12.

If a problem specifically asks for a ratio that compares a specified unit, then you must change all measurements to the specified unit. Otherwise, use the hint above to make it easier for you.

BRAIN TICKLERS
Set # 11

Solve each word problem with ratios by reading it carefully and thoroughly, planning a strategy, doing the plan, and then checking your work. Then, in your own words, list the steps you used to solve the problem.

1. A school recipe for brownies requires 6 cups of flour and 8 cups of sugar. This will yield enough brownies for the entire sixth grade. Find the ratio of cups of sugar to cups of flour.

2. A candy bar that cost 25¢ in 1980 costs 60¢ in 2000. Find the ratio of the *increase* in the candy bar's price to its cost in 1980.

3. A pair of running sneakers costs $70 and a pair of cross-training sneakers costs $84. Find the ratio of the cost of the running sneakers to the cost of the cross trainers.

4. A standard-sized bowl of spaghetti weighs 2 ounces. The biggest bowl of spaghetti ever made weighed 605 pounds. What is the ratio of the standard-sized bowl to the biggest bowl? (16 ounces = 1 pound)

5. Burger City sold 75 hamburgers between 12:30 PM and 1:00 PM, and 100 hamburgers between 1:00 PM and 1:30 PM. Find the ratio of the number of hamburgers sold between 12:30 PM and 1:00 PM to the number sold between 1:00 PM and 1:30 PM.

6. At the Slice of Life restaurant, prices for children under age 12 are $1.25 per slice of pizza, 80¢ for a small soda, and 45¢ for an ice cream cone.

   a. Find the ratio of the price of a slice of pizza to the price of an ice cream cone.

b. Find the ratio of the price of a slice of pizza to the total cost of one soda and one ice cream cone.

7. Hughes University offers a summer math course that meets five days a week over a two-week span for five hours per day. The same course, offered in the fall, meets once a week for ten weeks for two hours per class. Find the ratio of the total hours in the summer course to the total hours in the fall course.

(Answers are on page 130.)

Ratios lead to rates
For the next topic, so see
How word problems are painless
When they're read carefully.

WORD PROBLEMS WITH RATES

When you compare two quantities with different units, you are finding a *rate*.

Hints for finding rates

Look for the different units and their corresponding numbers. For example, look at the problem below.

Oranges cost $2.49 for a 3-pound bag. What is the price per pound of the oranges?

The units in this example are dollars and pounds. The rate is the number of dollars for one pound of oranges.
Create a fraction by placing the total amount in the numerator and the size of one group in the denominator. In this case, the total number is the total amount ($2.49) and the size of one group is the weight of one bag (3 pounds). The fraction is $\frac{\$2.49}{3\text{ lb}}$.

Simplify the fraction by dividing the numerator by the denominator.

$$\frac{\$2.49}{3\text{ lb}} = \$2.49 \div 3\text{ lb} = \$0.83 \text{ for one pound, or } \$0.83 \text{ per pound}$$

MATH NOTE

Look for the following words, which are often used in rate problems: *for, in, per, to, each.*

Here are some examples of rates: 55 miles *per* hour, $51 *for* 6 hours of work, 20 brownies *for* 5 people, 3 slices of pizza *in* 12 minutes, 200 people *to* a square mile, 2 pieces of fruit *for each* person.

EXAMPLE:

Shop-A-Lot Market advertised 5 pounds of potatoes for $2.55. At this rate, find the cost of one pound of potatoes.

5 lbs. POTATOES $2.55

Step 2: Plan a strategy.
Form the fraction with the total cost of 5 pounds of potatoes in the numerator and the total weight, 5 pounds, in the denominator.
Divide to find the cost of one pound of potatoes.

Step 3: Do the plan.

$$\frac{\$2.55}{5 \text{ lb}} = \$2.55 \div 5 \text{ lb.} = \$0.51 \text{ per pound of potatoes}$$

The price of one pound of potatoes (the *unit price*) is 51¢.

Step 4: Check your work.
Check that you have formed the correct fraction, with the total cost in the numerator and the number of pounds in the denominator.
Check your division by multiplication.

$$\$0.51/\text{lb} \times 5 \text{ lb} = \$2.55 \text{ for 5 pounds} \quad ✓$$

EXAMPLE:

To break a world record, a man once shaved 278 men in 60 minutes. At this rate, about how many men could he shave in one minute?

Step 2: Plan a strategy.
Form the fraction with the total number of men shaved in the numerator and the total time in minutes to shave them in the denominator.

Divide to find the number of men shaved in one minute.

Step 3: Do the plan.

$$\frac{278 \text{ men}}{60 \text{ min}} = 278 \text{ men} \div 60 \text{ min} \approx 4.63 \text{ men per minute.}$$

Since the question asks "about how many," round 4.63 to 5.

The record setter could shave about 5 men in one minute.

Step 4: Check your work.

Check that you have formed the correct fraction, with the total number of men in the numerator and the total number of minutes in the denominator.

Check your division with multiplication.

4.63 men shaved per minute × 60 minutes ≈
277.8 which rounds to 278 men ✓

EXAMPLE:

Celia bought eight six-packs of soda for a picnic. If 16 people were at the picnic, and each person was allotted the same number of cans of soda, how many cans did each person receive if all of the cans were distributed?

Step 2: Plan a strategy.

First multiply to find the number of cans of soda bought for the picnic.

Then form the fraction with the total number of cans of soda in the numerator and the number of people in the denominator.

Divide to find the number of cans of soda per person.

Step 3: Do the plan.

8 packs × 6 cans per pack = 48 cans of soda

$$\frac{48 \text{ cans}}{16 \text{ people}} = 48 \text{ cans} \div 16 \text{ people} = 3 \text{ cans of soda per person}$$

There were 3 cans of soda per person at the picnic.

Step 4: Check your work.

Check your multiplication with division.

48 cans of soda ÷ 8 packs = 6 cans of soda in a pack, or

48 cans of soda ÷ 6 cans in a pack = 8 packs of soda ✓

Check that you have formed the correct fraction, with the total number of cans of soda in the numerator and the number of people at the picnic in the denominator. Check your division by multiplication.

3 cans of soda per person × 16 people = 48 cans of soda ✓

BRAIN TICKLERS
Set # 12

Solve each word problem with rates by reading it carefully and thoroughly, planning a strategy, doing the plan, and then checking your work. Then, in your own words, list the steps you used to solve the problem.

1. Last week, Aamina worked seven hours each day Monday through Thursday, and eight hours on Saturday. She earned a total of $333. If her hourly rate was constant, find the amount of money that Aamina earned for each hour of work.

2. Shop-A-Lot Market is advertising 3 pounds of bananas for $2.37. Food Farm is advertising 2 pounds of bananas for $1.70. Which supermarket is advertising the least expensive price per pound for bananas?

3. Traveling cross-country, the Beeper family rode 510 miles in 8.5 hours. At this rate, how many miles did the Beepers drive per hour?

4. The most expensive meal ever purchased was by three diners in London, England. The diners spent close to $21,000 on the meal. If the diners shared the cost equally, what was the cost per person?

5. Mr. Scrub offers three ways to pay for car washes: a book of six car wash coupons for $33.00, a special offer of two washes for $11.50, or one wash for $5.95. Which option offers the least expensive unit price for one car wash?

6. Linda baked three pans of lemon bars with 24 bars per pan for a staff meeting. If 24 staff members were at the meeting and each person ate the same number of bars, with none left over, how many bars did each person eat?

7. For a read-a-thon, a fifth grade class read 243 books in 4.5 weeks. At this rate, how many books were read per week?

8. Perry Graph stacked his collection of 28 history books side by side on a bookshelf. The length of the books on the shelf measured 42 inches. At this rate, how many books could be placed per foot of bookshelf?

9. As part of her interval training, Rosie Runner jogs 1.5 miles in 15 minutes and then walks 2 miles in the next 30 minutes. What is the difference between her hourly rate for jogging and her hourly rate for walking?

10. In Connecticut, there is a tax charge of $0.84 on a $14 dinner bill. Find the tax rate in Connecticut. (Tax rates are based on cents per one dollar.)

(Answers are on page 131.)

Ratios and rates
Are just part of our plan
To solve all types of problems—
Here come proportions to our hand!

WORD PROBLEMS WITH PROPORTIONS

Follow the same steps to solve problems with proportions that you have used for solving ratio and rate problems.

MATH NOTE

When you compare two ratios or two rates that are equal in value, you are dealing with a *proportion*. (Two equivalent fractions can be called a proportion.)

Here are some examples of proportions:

$1.00 for 4 pounds = $0.25 for one pound or
$$\frac{1.00}{4} = \frac{0.25}{1}$$

44 miles per hour = 11 miles per 15 minutes or
$$\frac{44}{60} = \frac{11}{15}$$

You can check if two rates are equal, and therefore form a proportion. In a proportion, the cross products are equal.

$$\frac{1.00}{4} \quad \frac{0.25}{1}$$
The cross products are $1.00 \times 1 = 1.00$ and $4 \times 0.25 = 1.00$.

$$\frac{44}{60} \quad \frac{11}{15}$$
The cross products are $44 \times 15 = 660$ and $60 \times 11 = 660$.

Hints for working with proportions

The scale on a map shows that 2 inches represent an actual distance of 50 miles. The map-distance between two cities on the map is 7 inches. Sam Smart says that the actual distance between the two cities is 175 miles. Is Sam correct?

When you are comparing two rates, look for the two different units in each rate. In this case the units are inches and miles.

Create a fraction using words by placing one unit in the numerator and the other unit in the denominator. In this case you can create the fraction $\frac{\text{inches}}{\text{miles}}$. (It does not matter which of the two units is placed in the numerator. The other unit is placed in the denominator.)

Create two fractions by looking for the numbers that correspond to each unit in the fraction. Here the fractions are $\frac{2 \text{ inches}}{50 \text{ miles}}$ and $\frac{7 \text{ inches}}{175 \text{ miles}}$.

Set the two fractions equal to each other. (Proportions are two equal rates or fractions.)

$$\frac{2}{50} \overset{?}{=} \frac{7}{175}$$

(The question mark here means that we do not know yet whether the fractions are equal. That is the question!)

Use cross products to check if the two rates are equal.

$$2 \times 175 \overset{?}{=} 50 \times 7$$

$$350 = 350 \quad \checkmark$$

The two rates are equal. Thus, 7 inches corresponds to 175 miles. Sam is correct.

EXAMPLE:

Shop and Save	Star Shopper
Special!	Special!
Delicious Apples	Delicious Apples
4 pound bag for $2.36	6 pound bag for $3.54

Which store has the better buy on Delicious apples?

Step 2: Plan a strategy.

Determine the unit price of Delicious apples at each of the two stores.

Compare the unit prices to determine which store has the better buy, or if the unit prices are the same.

Step 3: Do the plan.

For Shop and Save Market, form the fraction with the total cost of 4 pounds of apples in the numerator and the total weight, 4 pounds, in the denominator. Divide to find the price for one pound of apples at Shop and Save Market.

$$\frac{\$2.36}{4 \text{ lb}} = \$2.36 \div 4 \text{ lb} = \$0.59 \text{ per pound at Shop and Save}$$

Now do the same for Star Shopper.

$$\frac{\$3.54}{6 \text{ lb}} = \$3.54 \div 6 \text{ lb} = \$0.59 \text{ per pound at Star Shopper}$$

Both stores have the same unit price. Neither store has a better buy. When two rates are equal, they form a proportion.

$$\frac{\$2.36}{4} = \frac{\$3.54}{6} \text{ is a proportion.}$$

Step 4: Check your work.

Check your divisions with multiplications.

$$\$0.59 \times 4 = \$2.36 \quad \checkmark$$

$$\$0.59 \times 6 = \$3.54 \quad \checkmark$$

You can also check with cross products to show that the rates are equal.

$$\frac{\$2.36}{4} \overset{?}{=} \frac{\$3.54}{6}$$

$$2.36 \times 6 \overset{?}{=} 4 \times 3.54$$

$$14.16 = 14.16 \quad \checkmark$$

EXAMPLE:

Florine has a cookie recipe that yields 36 cookies. The recipe requires $1\frac{1}{2}$ cups of flour. Florine wants to increase the ingredients to yield 48 cookies. How many cups of flour will she need?

Step 2: Plan a strategy.

The two different units are cups and cookies.

Set the $\dfrac{\text{cups of flour}}{\text{cookies}}$ rate for each recipe equal to each other to form a proportion.

MATH NOTE

If a unit in one part of a rate fraction does not have a corresponding number, use a symbol in its place. The symbol, usually a letter, represents the missing number. In this Example, we use the letter c.

Step 3: Do the plan.

The proportion is $\dfrac{1\frac{1}{2} \text{ cups of flour}}{36 \text{ cookies}} = \dfrac{c \text{ cups of flour}}{48 \text{ cookies}}$.

Cross multiply.

$$1\frac{1}{2} \times 48 = 36 \times c$$

$$72 = 36 \times c$$

Divide both sides by 36.

$$72 \div 36 = (36 \times c) \div 36$$

$$2 = c, \text{ or } c = 2 \text{ cups}$$

Two cups of flour will be needed to make 48 cookies.

Step 4: Check your work.

Substitute 2 cups for c in the proportion

$$\dfrac{1\frac{1}{2} \text{ cups of flour}}{36 \text{ cookies}} \stackrel{?}{=} \dfrac{2 \text{ cups of flour}}{48 \text{ cookies}}$$

Cross multiply to check if this is a proportion.

$$1\frac{1}{2} \times 48 \stackrel{?}{=} 36 \times 2$$

$$72 = 72 \ \checkmark$$

The answer, 2 cups, is correct.

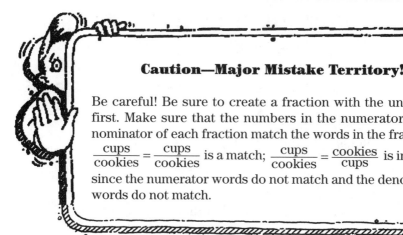

Caution—Major Mistake Territory!

Be careful! Be sure to create a fraction with the unit words first. Make sure that the numbers in the numerator and denominator of each fraction match the words in the fraction: $\frac{\text{cups}}{\text{cookies}} = \frac{\text{cups}}{\text{cookies}}$ is a match; $\frac{\text{cups}}{\text{cookies}} = \frac{\text{cookies}}{\text{cups}}$ is incorrect, since the numerator words do not match and the denominator words do not match.

EXAMPLE:

On a map, every 2 inches represents an actual distance of 100 miles. Find the actual distance between two towns if the map distance is 5 inches.

Step 2: Plan a strategy.

The two different units are

inches and miles.

Set the $\frac{\text{inches}}{\text{miles}}$ rate for the map

scale and the $\frac{\text{inches}}{\text{miles}}$ rate for the

distance between the towns equal to each other.

Since the actual distance between the towns is not

known, use a symbol in its place.

Step 3: Do the plan.

The proportion is $\frac{2 \text{ inches}}{100 \text{ miles}} = \frac{5 \text{ inches}}{m \text{ miles}}$.

Cross multiply.

$$2 \times m = 100 \times 5$$

$$2 \times m = 500$$

Divide both sides of the equation by 2.

$$(2 \times m) \div 2 = 500 \div 2$$

$$m = 250 \text{ miles}$$

The actual distance between the two towns is 250 miles.

Step 4: Check your work.

Substitute 250 miles for m in the proportion.

$$\frac{2 \text{ inches}}{100 \text{ miles}} \overset{?}{=} \frac{5 \text{ inches}}{250 \text{ miles}}$$

Cross multiply to check if this is a proportion.

$$2 \times 250 \overset{?}{=} 100 \times 5$$

$$500 = 500 \quad \checkmark$$

The answer, 250 miles, is correct.

MATH NOTE

When you are solving proportion problems, the variable may end up in either the numerator or the denominator of a fraction. It will depend on the unknown number and its place in matching with the units.

EXAMPLE:

In the seventh grade class, the ratio of boys to girls is 3 : 4. How many girls are in the class if there is a total of 140 students in the class?

Step 2: Plan a strategy.

Since we know the total number of students, and not the number of boys, we cannot use the ratio of 3 : 4, or 3 boys for every 4 girls, to find the number of girls directly.

Find the ratio of the number of girls to the total number of students.

Then write a proportion and solve for the number of girls in the class.

Step 3: Do the plan.

If there are 3 boys for every 4 girls, then there are 4 girls for every 3 boys + 4 girls = 7 students. Thus the ratio to use to find the number of girls in the class of 140 students is 4 girls for every 7 students, or 4 : 7. Write a proportion.

$$\frac{4 \text{ girls}}{7 \text{ students}} = \frac{g \text{ girls}}{140 \text{ students}}$$

Since the number of girls in the class is not known, the letter g is used to represent the missing number.

Cross multiply.

$$4 \times 140 = 7 \times g$$
$$560 = 7 \times g$$

Divide both sides of the equation by 7.

$$560 \div 7 = (7 \times g) \div 7$$
$$80 = g$$

There are 80 girls in the class of 140 students.

Step 4: Check your work.

Substitute 80 for g in the proportion.

$$\frac{4 \text{ girls}}{7 \text{ students}} \overset{?}{=} \frac{80 \text{ girls}}{140 \text{ students}}$$

Cross multiply to check if this is a proportion.

$$4 \times 140 \overset{?}{=} 7 \times 80$$
$$560 = 560 \quad ✓$$

The answer, 80 girls, is correct.

BRAIN TICKLERS
Set # 13

Solve each word problem by planning and carrying out a strategy. Remember to check your work. At the end of each problem you will find a letter. By placing this letter on the line above the corresponding answer in the answer code following the problems, you will be able to answer this question:

In what film did Mickey Mouse make his first screen appearance?

1. Rod Racer can run $\frac{3}{4}$ of a race in 18 minutes. At that rate, how long would it take him to run the whole race? (L)

2. To break a record, a group of students once pushed a bathtub on wheels close to 320 miles in 24 hours. At that rate, how many miles could the bathtub be pushed in 6 hours? (E)

3. Price Saver is advertising 3 pounds of oranges for $1.29. The unit price for oranges is the same at both Price Saver and at Cost Mart. If a bag of oranges costs $2.15 at Cost Mart, how many pounds does the bag weigh? (S)

4. At Camp Greenwood, the ratio of counselors to campers is 3 to 20. How many campers attend Camp Greenwood if there are 36 counselors? (O)

5. Find the actual distance between Twin Oaks and Triple Falls if the distance on the map is 3 inches. (W)

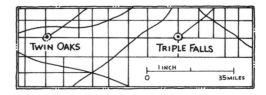

6. What would be the map distance between Unit City and Prime Town if the actual distance between these two cities is $81\frac{1}{4}$ miles, and the scale of the map is 3 inches = 25 miles. (T)

7. A photograph that measures $3\frac{1}{2}$ inches high and 5 inches wide was enlarged. The enlargement is $12\frac{1}{2}$ inches wide. If the height and width were enlarged in the same ratio, what is the height of the enlarged picture? (L)

8. The ratio of students studying French to students studying Spanish is 2 : 3. If there are 150 students studying one of either French or Spanish, how many students are studying Spanish? (A)

9. Bini Baser had 24 hits in 60 times at bat. At that rate, how many hits should Bini be expected to get in 100 times at bat? (I)

10. The ratio of boys to girls in the tennis league is 4 : 5. If there are 63 children altogether in the league, how many boys are in the tennis league? (M)

11. To break a record, a woman once typed 216 words in one minute. At that rate, how many words could she type in seven minutes? (A)

12. A picture that measures $8\frac{1}{2}$ inches high and 11 inches wide needs to be reduced so that it can fit into a frame that measures $6\frac{4}{5}$ inches high. If all measurements are reduced in the same ratio, what will be the width of the reduced picture? (T)

13. A CD, priced at $12, costs $12.60 with the tax included. At this tax rate, what will the tax be on a CD set priced at $20? (I)

14. The ratio of bagels to muffins sold on Monday at Sunshine Bakery was 5 : 2. How many bagels and muffins combined were sold on Monday if 120 muffins were sold? (B)

15. The tax rate in New York City is $8\frac{1}{4}$¢ per dollar. Find the total cost, including the tax, of a telephone whose price is $48. (E)

Answer Code

Mickey Mouse first appeared in

5	8.8	80	90	28	420	240	1,512	9.75

52.5	1.00	24	8.75	40	51.96

(Answers are on page 134.)

BRAIN TICKLERS—THE ANSWERS

Set #11, page 115

1. The ratio of the number of cups of sugar to the number of cups of flour is $\frac{8}{6}$, which reduces to $\frac{4}{3}$, or 4 to 3.

2. First find the increase in price from 1980 to 2000:
   60¢ − 25¢ = 35¢.
   The ratio of increase in price from 1980 to 2000 is $\frac{35¢}{25¢}$, which reduces to $\frac{7}{5}$, or 7 to 5.

3. The ratio is $\frac{\$70 \text{ running shoes}}{\$84 \text{ cross-training shoes}}$, which simplifies to $\frac{5}{6}$, or 5 to 6.

4. First change pounds to ounces.

   605 pounds × 16 ounces per pound = 9,680 ounces

   The ratio is $\frac{2 \text{ oz}}{9,680 \text{ oz}}$, which reduces to $\frac{1}{4,840}$, or 1 to 4,840.
   (You could change 2 ounces to pounds. However, this will yield a decimal or fraction in the numerator of the ratio, making it more difficult to reduce. The simplified ratio will still be 1 to 4,840.)

5. The ratio of the number of hamburgers sold between 12:30 PM and 1:30 PM to the number of hamburgers sold from 1:00 PM to

1:30 PM is $\dfrac{75 \text{ hamburgers}}{100 \text{ hamburgers}}$, which reduces to $\dfrac{3}{4}$, or 3 to 4.

6. a. First change the dollar cost of a slice of pizza to cents.

$$\$1.25 = \$1.25 \times 100 \text{ cents per dollar} = 125¢$$

The ratio of the price of a slice of pizza to the price of an ice cream is $\dfrac{125¢}{45¢}$, which reduces to $\dfrac{25}{9}$, or 25 to 9.

b. Find the total cost of one soda and one ice cream cone: $80¢ + 45¢ = 125¢$.

The ratio of the price of a slice of pizza to the total cost of one soda and one ice cream is $\dfrac{125¢}{125¢}$, which reduces to $\dfrac{1}{1}$, or 1 to 1.

7. First find the total number of hours in the summer course and the fall course.

Summer course: 5 days per week × 2 weeks = 10 days

10 days × 5 hours per day = 50 hours

Fall course: 1 day per week × 10 weeks = 10 days

10 days × 2 hours per day = 20 hours

The ratio of hours in the summer course to hours in the fall course is $\dfrac{50 \text{ h}}{20 \text{ h}}$, which reduces to $\dfrac{5}{2}$, or 5 to 2.

Set # 12, page 119

1. First find the total hours that Aamina worked last week. There are 4 days from Monday through Thursday. Aamina worked 7 hours each of these days. She also worked 8 hours on Saturday.

Monday through Thursday: 4 days × 7 hours per day = 28 hours

Saturday: 1 day × 8 hours per day = 8 hours

Total hours: 28 + 8 = 36 hours

Aamina earned a total of $333 for 36 hours of work.

Her hourly rate is $\dfrac{\$333}{36 \text{ hours}} = \$9.25$.

2. First find the unit price of bananas at each market.

   Shop-A-Lot: $\dfrac{\$2.37}{3 \text{ pounds}} = \$0.79$ per pound

   Food Farm: $\dfrac{\$1.70}{2 \text{ pounds}} = \$0.85$ per pound

   Since $0.79 is less than $0.85, Shop-A-Lot is offering the lesser of the two unit prices.

3. Form the fraction with the total number of miles driven in the numerator and the total number of hours driven in the denominator.

$$\dfrac{510 \text{ mi}}{8.5 \text{ h}} = 60.0 \text{ miles per hour}$$

4. Form the fraction with the total cost in the numerator and the number of diners in the denominator.

$$\dfrac{\$21,000}{3 \text{ people}} = \$7,000 \text{ per person (Wow!)}$$

5. First find the unit cost for each car wash option.

   Book of 6 car wash coupons: $\dfrac{\$33.00}{6 \text{ washes}} = \$5.50$ per car wash

   Two washes for $11.50: $\dfrac{\$11.50}{2 \text{ washes}} = \$5.75$ per car wash

   One wash alone: $5.95 per car wash

   Since $5.50 < 5.75 < 5.95$, the coupon book has the least expensive unit price per car wash of the three options.

6. First find the total number of lemon bars that Linda baked.

$$3 \text{ pans} \times 24 \text{ bars per pan} = 72 \text{ bars}$$

   Form the fraction with the total number of bars in the numerator and the number of staff members in the denominator.

$$\dfrac{72 \text{ bars}}{36 \text{ people}} \text{ or 3 bars per person}$$

7. Form the fraction with the number of books read in the numerator and the number of weeks in the denominator.

$$\frac{243 \text{ books}}{4.5 \text{ weeks}} = 54 \text{ books per week}$$

8. Since the problem asks for books per foot, change 42 inches to its corresponding number of feet.

$$\frac{42 \text{ inches}}{12 \text{ inches per foot}} = 3.5 \text{ feet}$$

Form the fraction with the number of books in the numerator and the total length of the books in the denominator.

$$\frac{28 \text{ books}}{3.5 \text{ feet}} = 8 \text{ books per foot}$$

9. First find Rosie's minute rate for jogging.

$$\frac{1.5 \text{ miles}}{15 \text{ minutes}} = 0.1 \text{ mile per minute}$$

Now find Rosie's hourly rate for jogging.

0.1 mile per minute × 60 minutes per hour = 6 miles per hour

Find Rosie's minute rate for walking.

$$\frac{2 \text{ miles}}{30 \text{ minutes}} = \frac{1}{15} \text{ mile per minute}$$

Now find Rosie's hourly rate for walking.

$\frac{1}{15}$ mile per minute × 60 minutes per hour = 4 miles per hour

(Note that the rate of $\frac{1}{15}$ mile per minute was not converted to its repeating decimal form 0.0666.... In this case it is easier to work with the fraction than the repeating decimal.)

10. Form the fraction with the tax charge in the numerator and the cost of the dinner in the denominator.

$$\frac{\$0.84}{\$14} = \$.06 \text{ per dollar, or } 6¢ \text{ per dollar}$$

Set #13, page 128

1. Since the rate is the same for $\frac{3}{4}$ of the race as for the whole race, write a proportion. Use t to represent the time in minutes to run the whole race.

$$\frac{\frac{3}{4} \text{ of the race}}{18 \text{ minutes}} = \frac{1 \text{ whole race}}{t \text{ minutes}}$$

Rewrite $\frac{3}{4}$ as the decimal 0.75 and cross multiply.

(The proportion can also be solved with the fraction in the first ratio.)

$$0.75 \times t = 18 \text{ minutes}$$

Divide both sides of the equation by 0.75.

$$t = 18 \div 0.75 = 24 \text{ minutes}$$

Write an L on the line above 24.

2. Write a proportion. Use m to represent the number of miles the tub could be pushed in 6 hours.

$$\frac{320 \text{ miles}}{24 \text{ hours}} = \frac{m \text{ miles}}{6 \text{ hours}}$$

Cross multiply.

$$320 \text{ miles} \times 6 \text{ hours} = 24 \text{ hours} \times m \text{ miles}$$

Divide both sides of the equation by 24.

$$m = (320 \times 6) \div 24 = 1920 \div 24 = 80 \text{ miles in six hours}$$

Write an E on the line above 80.

3. Since the unit prices at Price Saver and Cost Mart are equal, write a proportion. Use p to represent the number of pounds in a bag of oranges at Cost Mart.

$$\frac{\$1.29}{3 \text{ pounds}} = \frac{\$2.15}{p \text{ pounds}}$$

Cross multiply.

$$\$1.29 \times p \text{ pounds} = 3 \text{ pounds} \times \$2.15$$

Divide both sides of the equation by 1.29.

$$p = (3 \times 2.15) \div 1.29 = 6.45 \div 1.29 = 5 \text{ pounds}$$

Write an S on the line above 5.

4. Write a proportion. Use c to represent the number of campers if there are 36 counselors.

$$\frac{3 \text{ counselors}}{20 \text{ campers}} = \frac{36 \text{ counselors}}{c \text{ campers}}$$

Cross multiply.

$$3 \times c = 20 \times 36$$

Divide both sides of the equation by 3.

$$c = (20 \times 36) \div 3 = 720 \div 3 = 240 \text{ campers}$$

Write an O on the line above 240.

5. Write a proportion. Use m to represent the number of miles if the map distance is 3 inches.

$$\frac{2 \text{ inches}}{35 \text{ miles}} = \frac{3 \text{ inches}}{m \text{ miles}}$$

Cross multiply.

$$2 \times m = 35 \times 3$$

Divide both sides of the equation by 2.

$$m = (35 \times 3) \div 2 = 105 \div 2 = 52.5 \text{ miles}$$

Write a W on the line above 52.5.

6. Write a proportion. Use i to represent the number of inches when the actual distance is $81\frac{1}{4}$ miles.

$$\frac{3 \text{ inches}}{25 \text{ miles}} = \frac{i \text{ inches}}{81\frac{1}{4} \text{ miles}}$$

Write $81\frac{1}{4}$ as the decimal 81.25.

$$\frac{3 \text{ inches}}{25 \text{ miles}} = \frac{i \text{ inches}}{81.25 \text{ miles}}$$

Cross multiply.

$$3 \times 81.25 = 25 \times i$$

Divide both sides of the equation by 25.

$$i = (3 \times 81.25) \div 25 = 9.75 \text{ inches}$$

Write a T on the line above 9.75.

7. Write a proportion. Use i to represent the height in inches of the enlarged photograph.

$$\frac{3\frac{1}{2} \text{ inches high}}{5 \text{ inches wide}} = \frac{i \text{ inches high}}{12\frac{1}{2} \text{ inches wide}}$$

Write $3\frac{1}{2}$ and $12\frac{1}{2}$ as the decimals 3.5 and 12.5.

$$\frac{3.5 \text{ inches high}}{5 \text{ inches wide}} = \frac{i \text{ inches high}}{12.5 \text{ inches wide}}$$

Cross multiply.

$$3.5 \times 12.5 = 5 \times i$$

Divide both sides of the equation by 5.

$$i = (3.5 \times 12.5) \div 5 = 8.75 \text{ inches high}$$

Write an L on the line above 8.75.

8. You are asked to compare students studying only Spanish to the total number of students. You are given the rate of students studying French to students studying Spanish. In the ratio 2 : 3, the 3 stands for students studying Spanish. You

need to find the ratio of students studying Spanish to the total number of students. Add the numbers in the ratio, $2 + 3 = 5$, to find the proportional part of the whole represented by students who are studying Spanish. The result is a ratio of 3 students studying Spanish for each 5 students.

Write the proportion. Let s represent the number of students studying Spanish.

$$\frac{3 \text{ Spanish}}{5 \text{ Spanish or French}} = \frac{s \text{ Spanish}}{150 \text{ Spanish or French}}$$

Cross multiply.

$$3 \times 150 = 5 \times s$$

Divide both sides of the equation by 5.

$$s = 450 \div 5 = 90 \text{ students studying Spanish}$$

Write an A on the line above 90.

9. Write a proportion. Use h to represent the number of hits expected if Bini goes to bat 100 times.

$$\frac{24 \text{ hits}}{60 \text{ times at bat}} = \frac{h \text{ hits}}{100 \text{ times at bat}}$$

Cross multiply.

$$24 \times 100 = 60 \times h$$

Divide both sides of the equation by 60.

$$h = 2400 \div 60 = 40 \text{ hits with 100 times at bat}$$

Write an I on the line above 40.

10. The ratio of boys to girls is $4 : 5$. The ratio of boys to total children in the tennis league is $4 : (4 + 5) = 4 : 9$.
Write a proportion. Use b to represent the number of boys in the tennis league if there are 63 children in the league.

$$\frac{4 \text{ boys}}{9 \text{ children}} = \frac{b \text{ boys}}{63 \text{ children}}$$

Cross multiply.

$$4 \times 63 = 9 \times b$$

Divide both sides of the equation by 9.

$$m = 252 \div 9 = 28 \text{ boys in the tennis league}$$

Write an M on the line above 28.

11. Write a proportion. Let w stand for the number of words the typist could type in seven minutes.

$$\frac{216 \text{ words}}{1 \text{ minute}} = \frac{w \text{ words}}{7 \text{ minutes}}$$

Cross multiply.

$$216 \times 7 = 1 \times w$$

$$w = 1{,}512 \text{ words typed in seven minutes}$$

Write an A on the line above 1,512.

12. Write a proportion. Use w to represent the width in inches of the reduced picture.

$$\frac{8\frac{1}{2} \text{ inches high}}{11 \text{ inches wide}} = \frac{6\frac{4}{5} \text{ inches high}}{w \text{ inches wide}}$$

Write $8\frac{1}{2}$ and $6\frac{4}{5}$ as the decimals 8.5 and 6.8.

$$\frac{8.5 \text{ inches high}}{11 \text{ inches wide}} = \frac{6.8 \text{ inches high}}{w \text{ inches wide}}$$

Cross multiply.

$$8.5 \times w = 11 \times 6.8$$

Divide both sides of the equation by 8.5.

$$i = (11 \times 6.8) \div 8.5 = 74.8 \div 8.5 = 8.8 \text{ inches wide}$$

Write a T on the line above 8.8.

13. First find the amount of tax on the $12 purchase by subtracting $12.00 from $12.60. The tax is $0.60 on a $12 purchase. Write a proportion. Let t stand for the tax on the $20 CD set.

$$\frac{\$0.60 \text{ tax}}{\$12 \text{ purchase}} = \frac{t \text{ tax}}{\$20 \text{ purchase}}$$

Cross multiply.

$$0.60 \times 20 = 12 \times t$$

Divide both sides of the equation by 12.

$$t = (0.60 \times 20) \div 12 = 12 \div 12 = \$1.00 \text{ in tax on the \$20 CD set}$$

Write an I on the line above 1.00.

14. First find the ratio of muffins to bagels and muffins combined. The ratio of bagels to muffins is $5 : 2$. The ratio of muffins to bagels is $2 : 5$. Therefore, the ratio of muffins to bagels and muffins combined is $2 : (2 + 5) = 2 : 7$. Write a proportion. Use t to represent the number of bagels and muffins combined sold when 120 muffins were sold.

$$\frac{2 \text{ muffins}}{7 \text{ bagels and muffins}} = \frac{120 \text{ muffins}}{t \text{ bagels and muffins}}$$

Cross multiply.

$$2 \times t = 7 \times 120$$

Divide each side of the equation by 2.

$$t = (7 \times 120) \div 2 = 420 \text{ bagels and muffins sold}$$

Write a B on the line above 420.

15. First find the amount of the tax on a $48 telephone purchase. Change the dollar amounts to cents. Write a proportion. Use t to represent the amount of the tax on the telephone purchase.

$$\frac{8\frac{1}{4}\cent}{100\cent} = \frac{t\cent}{4800\cent}$$

Cross multiply.

$$8\frac{1}{4} \times 4800 = 100 \times t$$
$$39,600 = 100 \times t$$

Divide both sides of the equation by 100.

$t = 396¢ = \$3.96$ tax on the purchase of a \$48 telephone

Now find the total cost.

$\$48.00 + \$3.96 = \$51.96$ total cost of the telephone and tax

Write an E on the line above 51.96.

Answer Code

S	T	E	A	M	B	O	A	T
5	8.8	80	90	28	420	240	1,512	9.75

W	I	L	L	I	E
52.5	1.00	24	8.75	40	51.96

Mickey Mouse made his first screen appearance in *Steamboat Willie*.

CHAPTER FIVE

Perky Percents

We've learned about ratios,
Decimals and fractions, too;
Now look at percents
And their importance to you!

WORD PROBLEMS WITH PERCENTS

To solve word problems with percents, you will again use Polya's four steps. Remember, before you can plan a strategy, you must understand the problem. To understand a problem with percents, you must first read the problem carefully. Then ask yourself these questions: What is the problem about? What are you being asked to find? Is there enough information?

You must also plan a strategy, do your plan, and check your answer.

MATH NOTE

A *percent* is a ratio that compares a number to 100. *Per* means *out of* and *cent* means *100*. So, 15% means 15 out of 100, 2% means 2 out of 100, and so on.

Percents are used in sports, in sales, by restaurants, by banks, and more. Percents can also be used to compare numbers that are not out of 100. Let's see how we can find our new friend, percent.

Hints for finding percents

One way to find a percent is to use a proportion.

What percent of 120 is 24?

First form a ratio. The total or whole is the denominator of the ratio, in this case, 120. The numerator is the part, or piece, of the total. In this case, the numerator is 24, so the ratio is $\frac{24}{120}$. Now form a ratio with a denominator of 100 (100 is the whole) and a symbol in the numerator to represent the unknown percent: $\frac{p}{100}$. Form a proportion.

$$\frac{24}{120} = \frac{p}{100}$$

Cross multiply.

$$2400 = 120 \times p$$

Divide both sides of the equation by 120.

$$2400 \div 120 = p$$
$$20 = p$$
$$p = 20$$

This means that 24 out of 120 is the same as 20 out of 100, or 20%.

Another way to find a percent is to divide.

To find what percent of 120 that 24 is, again form the ratio of the part to the whole: $\frac{24}{120}$.

Divide the numerator by the denominator to find the corresponding decimal value.

$$24 \div 120 = 0.20$$

Multiply the decimal quotient by 100.

$$0.2 \times 100 = 20$$

Put a percent sign (%) on the product.

$$24 \text{ is } 20\% \text{ of } 120.$$

Many people who use calculators use this way more often because of the ease of performing long division with a calculator. Both ways can be used with pencil and paper.

EXAMPLE:
In a high school survey, it was found that 240 of 600 students walk to school. What percent of the students walk to school?

MATH NOTE

The word *of* frequently appears in percent problems. The number after the word *of* represents the total or whole and is the denominator of the ratio. (In this Example, *of the students* means of the total of 600 students. Thus, 600 will be the denominator.)

Step 2: Plan a strategy.

First form the ratio of the number of students who walk to school to the total number of students (part to whole). Now form a ratio with a denominator of 100 (100 is the whole) and a symbol in the numerator to represent the unknown percent.

Form and solve a proportion.

Step 3: Do the plan.

Be careful to put the numbers in the correct part of the ratio. In this case, one ratio is $\frac{240}{600}$, which reduces to $\frac{2}{5}$, and the other ratio is $\frac{p}{100}$.

Write and solve a proportion.

$$\frac{2}{5} = \frac{p}{100}$$

$$2 \times 100 = 5 \times p$$

$$p = 40$$

40 out of 100 students walk to school, so 40% of the students walk to school.

Step 4: Check your work.

First check that you have the numbers in the right order in the ratios and that you have reduced the ratio correctly. Substitute 40 for p and cross multiply to check that your proportion is correct.

$$\frac{2}{5} \overset{?}{=} \frac{40}{100}$$

$$2 \times 100 \overset{?}{=} 5 \times 40$$

$$200 = 200 \quad \checkmark$$

EXAMPLE:

There are 650 muscles in the human body. It takes 17 muscles to smile. To the nearest percent, what percent of all your muscles are needed to smile?

Step 2: Plan a strategy.
Use the division method.
First form the ratio of the number of muscles needed to smile to the total number of muscles (part to whole).
Divide and convert the resulting decimal to percent.

Step 3: Do the plan.
Be careful to put the numbers in the correct part of the ratio. In this case, the ratio of muscles for smiling to total muscles is $\frac{17}{650}$.
Divide 17 by 650.

$$17 \div 650 = 0.0261..., \text{ or about } 0.026$$

Multiply the decimal by 100.

$$0.026 \times 100 = 2.6$$

$$2.6\% \text{ is about } 3\%.$$

About 3% of the muscles in the human body are needed to smile.

Step 4: Check your work.
First check that you have the numbers in the right order in the ratio.
Then check your division and multiplication.

PERCENT CHANGE: INCREASE OR DECREASE

When there is a change in a quantity, often we are interested in the *percent of change*.
This change might be an increase or a decrease in the quantity.

Hints for finding percent of increase or decrease

What is the percent increase from 16 to 28?

First subtract to find the amount of increase.

$$28 - 16 = 12$$

Now form the ratio of the increase to the original amount. The original amount is the amount before the change.

$$\frac{\text{increase}}{\text{original amount}} = \frac{12}{16}$$

Divide.

$$12 \div 16 = 0.75$$

Write the decimal as a percent.

$$0.75 \times 100 = 75, \text{ so } 0.75 = 75\%$$

There is a 75% increase from 16 to 28.
You may also use a proportion to find the percent.

$$\frac{12}{16} = \frac{p}{100}$$

$$1200 = 16 \times p$$

$$p = 75$$

So, $\frac{12}{16} = 75\%$.

Find the percent decrease from 80 to 64.
First subtract to find the amount of decrease.

$$80 - 64 = 16$$

Now form the ratio of the decrease to the original amount. The original amount is the amount before the change.

$$\frac{\text{decrease}}{\text{original amount}} = \frac{16}{80}$$

Divide.

$$16 \div 80 = 0.2$$

Write the decimal as a percent.

$$0.2 \times 100 = 20, \text{ so } 0.2 = 20\%$$

There is a 20% decrease from 80 to 64.

EXAMPLE:

A $60 pair of shoes is on sale for $42. Find the percent of discount.

Step 2: Plan a strategy.
First find the decrease in price. Then find the ratio of the decrease in price to the original price.
Change the ratio to a percent.

Step 3: Do the plan.
Find the decrease in price.

$$\$60 - \$42 = \$18$$

Form the ratio of the decrease in price to the original price.

$$\frac{\text{decrease in price}}{\text{original amount}} = \frac{18}{60}$$

Change the ratio to a percent by dividing, multiplying by 100, and adding a percent sign.

$$18 \div 60 = 0.3 = 30\%$$

There is a 30% decrease (discount) in the price of the shoes.

Step 4: Check your work.
Work backwards.
Check your division with multiplication.

$$0.3 \times \$60 = \$18 \text{ discount on a pair of shoes} \quad \checkmark$$

Check your subtraction with addition.

$$\$18 + \$42 = \$60, \text{ the original price of the shoes} \quad \checkmark$$

Caution—Major Mistake Territory!

Be careful! Make sure that you use the *change* in the amount when you form a ratio to find a percent of increase or decrease. In the preceeding Example, you don't use 42 in the numerator to form the ratio $\frac{42}{60}$. This would give a result of 70%, which is the percent describing the ratio of the sale price to the original price of the shoes. The value $42 stands for the amount that was *paid*. The problem asks for the percent of discount, which describes the amount that was *saved*.

MATH NOTE

Percent changes can be greater than 100. This happens when the amount of change is more than the starting amount, that is, if the end number is more than twice the starting number. A ratio in which the numerator is greater than the denominator will equal a percent that is greater than 100%.

EXAMPLE:

Last year, Cara Caker invited 10 children to her birthday party. This year, she invited 35 children. What is the percent increase from the number invited to last year's party to the number invited to this year's party?

Step 2: Plan a strategy.
First find the increase in the number of children invited from last year to this year. Then find the ratio of the increase to the number invited last year.
Change the ratio to a percent.

Step 3: Do the plan.
Find the increase in the number of students invited.

$$35 - 10 = 25$$

Form the ratio of the increase in number to the number invited last year.

$$\frac{\text{increase}}{\text{number last year}} = \frac{25}{10}$$

Change the ratio to a percent by dividing, multiplying by 100, and adding a percent sign.

$$25 \div 10 = 2.5 = 250\%$$

There was a 250% increase in the number of children invited from last year to this year. (Remember that you may also use a proportion to find the percent.)

Step 4:   Check your work.
Work backwards.
Check your division with multiplication.

$2.5 \times 10 = 25$ more children invited this year than last ✓

Check your subtraction with addition.

$25 + 10 = 35$, the number of children invited this year ✓

Percent of discount or decrease and percent of increase or inflation are used in business, in shopping, and in population studies, to name just a few applications.

BRAIN TICKLERS
Set # 14

Solve each word problem for the percent or percent change. Read the problem carefully and thoroughly, plan a strategy, do the plan, and check your work.

1. Of 136 people visiting the art gallery, 17 bought paintings. What percent of the visitors bought a painting?

2. In Massachusetts there is a $0.60 tax on a purchase of $12. Find the percent of tax in Massachusetts.

3. There are 206 bones in the human body. There are 33 bones in the spinal column. To the nearest percent, what percent of your bones are in your spine?

4. Between 1924 and 1998, the United States won a total of 161 medals in the Winter Olympic Games. If the United States won 59 silver and 42 bronze medals, about what percent of its medals were gold medals?

5. A stereo, regularly priced at $240, is on sale for $192. With tax, the final price is $203.52.

   a. Find the percent discount.

   b. Find the percent tax rate.

6.
Nutritional Data for McDonald's

Item	Calories	Grams of Fat (1 gram of fat = 9 calories)
Hamburger	260	9
Cheeseburger	320	13
Big Mac	560	31

    a. For each item, what percent of its calories to the nearest percent are fat calories?

    b. Order the items from highest percent of fat to lowest percent of fat.

7. **Little League Enrollment**

Year	Boys	Girls
1998	120	80
2000	100	120

    a. In 1998, what percent of the children in the Little League were boys?

    b. Find the percent of increase in the number of girls from 1998 to 2000.

8.

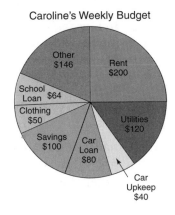

Caroline's Weekly Budget

    a. What percent of Caroline's total weekly budget is allocated for utilities?

    b. Which part of Caroline's budget represent 8% of her total weekly salary?

    c. Which part of the graph stands for 25% of Caroline's budget?

(Answers are on page 168.)

FIND THE WHOLE OR PART WHEN THE PERCENT IS KNOWN

Using ratios and proportions
To find percents, we have shown.
Now we find the part or the whole
When the percent is known.

There are two ways to solve a percent problem when the part or whole is not known. Learn both ways. Choose the one that you like best for the problem to be solved. Following is a problem solved using both plans.

What number is 35% of 7?

Plan One

1. Form a ratio, as before, with the total (whole) as the denominator and the unknown part as the numerator. In this case, the whole is 7.

    $$\frac{x}{7} \quad (x \text{ represents the part of the whole.})$$

2. Form a ratio with the known percent as the numerator and 100 as the denominator.

    $$\frac{35}{100}$$

3. Form a proportion and solve.

    $$\frac{x}{7} = \frac{35}{100}$$

    $$x \times 100 = 7 \times 35$$

    $$x \times 100 = 245$$

    $$x = 245 \div 100$$

    $$x = 2.45$$

35% of 7 is 2.45.

153

Plan Two

1. Instead of a proportion, use the equation Part = Percent × Whole.

$$\text{Part} = 35\% \times 7$$

2. Change the percent to a ratio. (Remember that a percent is a ratio that compares a number to 100.)

$$35\% = \frac{35}{100}$$

3. Divide to change the ratio to a decimal.

$$35 \div 100 = 0.35$$

4. Multiply the decimal by the total (whole).

$$0.35 \times 7 = 2.45$$

35% of 7 is 2.45.

MATH NOTE

Always change a percent to a decimal or fraction before performing a computation: $35\% \times 7$ is not 35×7, but 0.35×7.

Whichever plan you choose, percents will be painless. You cannot lose!

EXAMPLE:

A banana is about 76% water. How many ounces of water are in a banana that weighs 3 ounces?

Step 2: Plan a strategy.
Use Plan Two: Part = Percent × Whole.
Change the percent to a decimal.

Step 3: Do the plan.
Part = $76\% \times 3 = 0.76 \times 3 = 2.28$
About 2.28 ounces of a 3-ounce banana is water.

Step 4: Check your work.
Find what percent of 3 that 2.28 is.
Form the ratio $\frac{2.28}{3}$.

Divide to find the decimal: $2.28 \div 3 = 0.76$.
Change the decimal to a percent: $0.76 = 76\%$.
76% of a banana's weight is water. ✓

EXAMPLE:

The M&M/Mars Company states that about 30% of a one-pound bag of plain M&M candies are brown. If a bag has 150 brown candies, about how many total candies are in the bag?

Step 2: Plan a strategy.
Use Plan Two: Part = Percent × Whole.
Change the percent to a decimal.
Solve for the whole.

Step 3: Do the plan.

$$150 \text{ brown candies} = 30\% \times t$$

$$150 = 0.30 \times t$$

(t represents the total number of candies in a one-pound bag.)

Divide both sides of the equation by 0.30.

$$t = 500$$

There are about 500 candies in a one-pound bag of M&M candies.

Step 4: Check your work.
Find 30% of 500. This is the number of brown candies.
$30\% \times 500 = 0.30 \times 500 = 150$ brown candies in the bag ✓

You could also check by using Plan One.
Form a ratio with the number of brown candies as the numerator and the unknown total as the denominator: $\frac{150}{t}$.
Form a ratio with the percent number in the numerator and 100 in the denominator: $\frac{30}{100}$.

Write a proportion and cross multiply to solve for t.

$$\frac{150}{t} = \frac{30}{100}$$

$$150 \times 100 = 30 \times t$$

$$15{,}000 = 30 \times t$$

$$15{,}000 \div 30 = t$$

$$t = 500 \text{ candies per one-pound bag} \quad \checkmark$$

The answer is the same for the two different plans.

MATH NOTE

When you use Plan Two, you multiply when looking for the part (Part = Percent × Whole), and divide when looking for the whole (Whole = Part ÷ Percent).

Also, the percent must be written as a decimal or a fraction before performing any computations.

EXAMPLE:

A CD player is on sale for $12\frac{1}{2}$% off the regular price of $240.

Find the sale price of the CD player.

MATH NOTE

Percents that contain fractions can be tricky! But watch and see how painless it can be! Write the fraction in the percent as a decimal. That is, $12\frac{1}{2} = 12.5$. Now $12\frac{1}{2}\% = 12.5\%$. To change 12.5% to a decimal number, skip two places to the left of the decimal point in the percent to place the decimal point in the decimal form of the percent: $12\frac{1}{2}\% = 12.5\% = 0.125$. Be careful: $\frac{1}{2}\% = 0.5\% = 0.005$ (not 0.5).

Step 2: Plan a strategy.
Use Plan Two, Part = Percent × Whole, to find the amount of money that is saved.
Change the percent to a decimal.
To find the sale price, subtract the amount saved from the regular price.

Step 3: Do the plan.

$s = 12\frac{1}{2}\% \times \$240$, where s represents the amount saved

$s = 0.125 \times \$240 = \$30$ saved

Subtract the amount saved from the regular price to find the sale price.
Sale Price = Regular Price − Amount Saved = $240 − $30 = $210
The sale price of the CD player is $210.

Step 4: Check your work.
Work backwards.
Check your subtraction by addition.

$210 + $30 = $240, the regular price of the CD player ✓

Check your multiplication by division.

$30 ÷ $240 = 0.125 = 12.5%, the percent off the regular price, or

$30 ÷ 0.125 = $240, the regular price of the CD player ✓

PERCENT, FRACTION, AND DECIMAL EQUIVALENTS

Does $12\frac{1}{2}\%$ seem like a strange percent? There are some percents that appear often in percent problems. One of them is $12\frac{1}{2}\%$.

Some others are 20%, 25%, $33\frac{1}{3}\%$, and 50%. For these percents,

PERKY PERCENTS

using their equivalent fractions rather than their decimal forms makes the computations easier. The table below gives the percent, fraction, and decimal equivalents of some commonly occurring percents.

Percent	Fraction	Decimal
$12\frac{1}{2}\%$	$\frac{1}{8}$	0.125
20%	$\frac{1}{5}$	0.2
25%	$\frac{1}{4}$	0.25
$33\frac{1}{3}\%$	$\frac{1}{3}$	0.333...
50%	$\frac{1}{2}$	0.5

Thus, the CD player problem in the previous Example could be solved like this:

$s = 12\frac{1}{2}\% \times \$240$, where s represents the amount saved

$s = \frac{1}{8} \times \$240 = \$30$, the amount saved

$\$240 - \$30 = \$210$, the sale price

Whether a fraction or a decimal is used, through a percent problem you will cruise!

EXAMPLE:
A bicycle lock is on sale for 10% off of the regular price of $34.00. Find the amount of discount.

Step 2: Plan a strategy.
Use Plan Two, Part = Percent × Whole, where the part is the amount of discount.
Change the percent to a decimal.

Step 3: Do the plan.

$d = 10\% \times \$34$, where d represents the amount saved

$d = 0.10 \times \$34 = \$3.40$, the amount of discount

Step 4: Check your work.

Work backwards.

Check your multiplication by division.

$3.40 ÷ $34 = 0.10 = 10%, the discount percent, or

$3.40 ÷ 0.10 = $34, the regular price of the lock ✓

Some other commonly used percents are 5%, 10%, and 15%. There are ways to simplify computations with these percents. These are explained in the Hints below.

Hint for finding 10% of a number

When finding 10% of a number, move the decimal point of the number one place to the left.

What is 10% of 34?

Move the decimal point of 34, or 34., one place to the left.

10% of 34. = 3.4

Go back and look at the solution to the previous example. Using the "move the decimal point" hint, 10% × $34.00 = $3.40. Of course, the hint works because 10% is equivalent to the decimal 0.1.

Hint for finding 5% of a number

First find 10% of the number.
Then find half of the resulting number.

What is 5% of 56?

10% of 56 = 5.6

5.6 ÷ 2 = 2.8

2.8 is 5% of 56.

Check: 5% of 56 = 0.05 × 56 = 2.8. ✓

Notice that the two steps in the hint, move the decimal point and divide by two, may be done without paper and pencil!

Hint for finding 15% of a number

First find 10% of the number.
Then find 5% of the number.
Add the numbers for 10% and 5% of the number.

What is 15% of 40?

10% of 40 = 4

5% of 40 = 4 ÷ 2 = 2

4 + 2 = 6

15% of 40 = 6

Check: 15% of 40 = 0.15 × 40 = 6 ✓

EXAMPLE:

The Wilson's dinner bill at Debby's Delight totaled $20.60. They added a 15% tip for the server. What was the amount of the tip?

Step 2: Plan a strategy.
First find 10% of the dinner bill.
Then find 5% of the dinner bill.
Add the two amounts.

Step 3: Do the plan.

10% of $20.60 = $2.06

5% of $20.60 = $2.06 ÷ 2 = $1.03

total tip = $2.06 + $1.03 = $3.09

The tip on the Wilson's dinner bill was $3.09.

Step 4: Check your work.

15% of $20.60 = 0.15 × $20.60 = $3.09 ✓

Note that a tip of exactly 15% gives a tip amount of $3.09. Many people suggest a range of 15–20% for tipping at a restaurant. The Wilsons might have chosen to leave, for example, $3 (about 15%)

for satisfactory service, $4 (about 20%) for very good service, or $5 (about 25%) for exemplary service.

BRAIN TICKLERS
Set # 15

Solve each word problem. Read the problem carefully and thoroughly, plan a strategy, do the plan, and check your work. You may use Plan One, Plan Two, the 10%, 5%, or 15% hints, or change a percent to a fraction. The choice is yours!

The solution box following the problems contains the answers to the problems. As you answer each problem, cross out the answer in the solution box. Use the number that is not crossed out to answer the following question:

What percent of the human body is water?

1. In a school election, 48% of the 200 voters voted for Tara. How many people voted for Tara?

2. Carmine Cruiser has saved $630 toward the cost of a trip. This is 75% of the total cost of the trip. How much more money does Carmine need to save for the trip?

3. 144 fifth graders will be going to East Middle School next year. This is about 45% of all of the fifth graders. How many students are currently in the fifth grade?

4. A pair of sneakers, regularly priced at $84, is on sale for 20% off. Find the sale price.

5. Shopper's Paradise offers the following discount plan on items that remain on sale in the store for one week or more.

Week	Discount
1	10% off
2	15% off of Week 1 price
3	20% off of Week 2 price
4	25% off of Week 3 price
5	50% off of Week 4 price

What would be the cost of a suit regularly priced at $160 that has been on sale at shopper's paradise for three weeks?

6. This year, 45 kids signed up for the track and field team. This was 150% of the number of kids who were on the team last year. How many kids were on last year's team?

7. A dinner bill at Sid's Spaghetti Spot was $24.00. Find the amount of a 15% tip.

8. Holden sold $33\frac{1}{3}$% of his 42 raffle tickets. Lily sold 25% of her 60 raffle tickets. Lucinda sold $12\frac{1}{2}$% of her 72 raffle tickets. How many tickets did the person who sold the most tickets sell?

9. Students at the Miller Middle School try to sell 10% more at each monthly bake sale. If $50 worth of baked goods were sold in March, how much would the students need to sell in June to reach their goal?

Use the graph below to answer Questions 10 and 11.

Baldwin School Baseball Budget

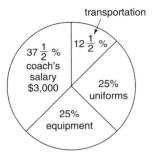

10. How much money is allocated for baseball uniforms?

11. How much money is allocated for transportation?

12. The price of a gallon of gasoline has increased 125% in two years. If the cost two years ago was $0.96 per gallon, what is the current price of gasoline?

Solution Box

96		174
320	67.20	97.92
30	3.60	15
66.55	2,000	2.16
	1,000	70

(Answers are on page 172.)

AN INTEREST IN INTEREST

Percents are used by banks to determine interest earned on savings and interest owed on loans. Let's look at some examples to see how money in savings accounts can grow!

There are two types of interest plans on savings that a bank can offer its customers: *simple interest* and *compound interest*.

Simple interest

In a *simple interest* account, you receive a set amount of interest once every year. If you leave your money in the account for less than a full year, the full-year interest is calculated, and you then receive the appropriate fractional part.

MATH NOTE

Many banks no longer offer simple interest. However, simple interest problems appear in most math books to provide a start for understanding compound interest.

Thus, it is a good idea to learn how to calculate simple interest first.

Following is an easy-to-use formula for finding simple interest. The *principal* is the original amount in the account.

Interest earned = Principal × Interest Rate (% per year) × Time (years)

or

$$I = P \times R \times T$$

Here is a sample problem using simple interest:
The annual simple interest on an $800 savings account is 4.5%. What is the interest earned after two years?

The original amount in the account, or principal, is $800. The interest rate, always stated as a percent, is 4.5% annually. The time is 2 years. The interest is unknown. Use the formula $I = P \times R \times T$. In this case, P = $800, R = 4.5% per year, and T = 2 years.

$$I = \$800 \times 4.5\% \times 2$$

Change the percent to a decimal: 4.5% = 0.045.
Solve.

$$I = \$800 \times 0.045 \times 2$$

$$I = \$72$$

The interest earned in two years is $72.

EXAMPLE:

Serelle Cently opened an account in a bank that offers a 5% simple annual interest rate. She deposited $500 into the account. If she doesn't deposit any additional money into the account, how much money will be in her account at the end of 4 years?

Step 2: Plan a strategy.

First use the formula $I = P \times R \times T$ to find the interest paid. Remember to change the percent to a decimal. Then find the total amount by adding the interest to the principal.

Step 3: Do the plan.
In this case, $P = \$500$, $R = 5\%$ per year, and $T = 4$ years.

$$I = \$500 \times 0.05 \times 4$$

$$I = \$100$$

The total amount Serelle will have in her account is $\$500 + \$100 = \$600$.

Step 4: Check your work.
Go back and check each of your calculations.

EXAMPLE:

Boris Biller invested $\$1,000$ in a money market account that offers a 6% simple annual interest rate. If he withdraws his money after six months, how much money in interest will he receive?

Step 2: Plan a strategy.
First use the formula $I = P \times R \times T$ to find the interest paid. Remember to change the percent to a decimal and remember to change the months to a fraction of a year.

Step 3: Do the plan.
In this case, $P = \$1,000$, $R = 6\%$ per year, and $T = 6$ months. So, $T = \frac{6}{12}$ year $= \frac{1}{2}$ year.

$$I = \$1,000 \times 0.06 \times \frac{1}{2}$$

$$I = \$30$$

Boris will receive $\$30$ in interest after six months.

Step 4: Check your work.
Go back and check each of your calculations.

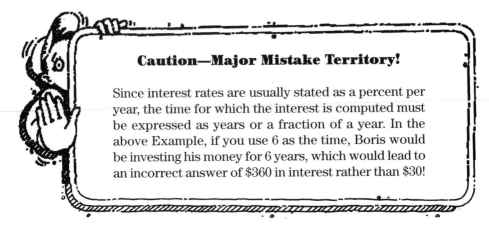

Caution—Major Mistake Territory!

Since interest rates are usually stated as a percent per year, the time for which the interest is computed must be expressed as years or a fraction of a year. In the above Example, if you use 6 as the time, Boris would be investing his money for 6 years, which would lead to an incorrect answer of $360 in interest rather than $30!

Compound interest

In a *compound interest* account, interest is added to your account multiple times per year, and your interest also earns interest. Some banks offer monthly compounded interest, some quarterly compounded interest (four times per year), and others more frequently compounded interest. There are many ways to compute compound interest, and banks now use computers to calculate the interest over a number of time periods. We shall just look at one way for computing compound interest based on repeated use of the simple interest formula.

As an illustration, consider the following problem.
A bank offers a savings account that earns a 5% interest rate compounded monthly. Find the amount of money after two months in an account with an initial principal of $1,000.

In this case, the principal is $1,000, the rate is 5%, and the time is two months.
One way to find the amount in the account is first to find the simple interest for one year. Then calculate the total amount in the account after one month. Do the same for month two based on the balance after month one.
Find the simple interest earned in one year.

$$I = P \times R \times T$$

$$I = \$1,000 \times 0.05 \times 1$$

$$I = \$50$$

This means that after one month $\left(\frac{1}{12} \text{ of a year}\right)$, the interest earned would be

$50 a year ÷ 12 months per year ≈ $4.17.

The amount in the account at the end of the first month is now

Amount = $1,000 + $4.17 = $1,004.17.

Now use this amount as the new principal to calculate the interest earned the second month. Find the simple interest on this new principal for one year.

$$I = P \times R \times T$$

$$I = \$1,004.17 \times 0.05 \times 1$$

$$I \approx \$50.21$$

This means that during the second month $\left(\frac{1}{12} \text{ of a year}\right)$, the interest earned would be

$50.21 a year ÷ 12 months per year ≈ $4.18.

The amount at the end of the second month is now

Amount = $1,004.17 + $4.18 = $1,008.35.

This amount is not much different from the amount with simple interest for two months ($1,008.33). However, if money is left in an account that compounds monthly over a longer period of time, much more interest is earned by compounding than with simple interest.

BRAIN TICKLERS
Set # 16

1. Mariah Monies invested $2,000 in a bank offering a 6% simple yearly interest rate. After one year, how much money was in her account?

2. Dorian McDime invested $600 for 3 months in an account offering a 4% simple interest rate. How much money did he withdraw at the end of the three months?

3. Little Town Bank offers a 5% interest rate, compounded monthly. On June 1, Lyle Lira deposited $500 in Little Town Bank. By August 1, how much money was in Lyle's account?

4. After one year, Kelly Counter received $55 in interest from her initial investment of $1,000. What simple interest rate did her bank offer?

5. Penelope Pound earned $45 in interest for an investment of $1,500 at a 6% simple annual interest rate. For how long was the money invested?

(Answers are on page 176.)

BRAIN TICKLERS—THE ANSWERS

Set # 14, page 151

You may have used another plan than that shown in an answer. That is fine.

1. 136 is the total (whole) and 17 is the part. Write and solve a proportion.

$$\frac{17}{136} = \frac{x}{100}$$

$$17 \times 100 = 136 \times x$$

$$1{,}700 = 136 \times x$$

$$1{,}700 \div 136 = x$$

$$x = 12\tfrac{1}{2}\%$$

$12\tfrac{1}{2}\%$ of the visitors bought a painting.

2. $12 is the whole and $0.60 is the part. The ratio is $\frac{0.60}{12}$. Change the ratio to a percent by dividing, multiplying by 100, and adding a percent sign.

$$0.60 \div 12 = 0.05 = 5\%$$

The tax rate in Massachusetts is 5%.

3. 206 is the whole and 33 is the part. Write and solve a proportion.

$$\frac{33}{206} = \frac{x}{100}$$

$$33 \times 100 = 206 \times x$$

$$3{,}300 = 206 \times x$$

$$3{,}300 \div 206 = x$$

$$x \approx 16.0$$

About 16% of your bones are in your spine.

4. There were 161 total medals awarded. The part, the number of gold medals, is not stated. Find the part that is not gold by adding the number of silver medals (59) and the number of bronze medals (42): 59 + 42 = 101 silver and bronze medals. Subtract 101 from 161 to get 60 gold medals. Form a ratio where 60 is the part and 161 is the total.

$$\frac{60}{161} = 60 \div 161 \approx 0.37, \text{ or } 37\%.$$

37% of the medals were gold.

5. a. The amount of discount is $240 (regular price) – $192 (sale price) = $48. Use the ratio $\dfrac{\text{discount}}{\text{regular price}}$, or $\dfrac{48}{240}$ to write and solve a proportion.

$$\frac{48}{240} = \frac{x}{100}$$

$$48 \times 100 = 240 \times x$$

$$4,800 = 240 \times x$$

$$4,800 \div 240 = x$$

$$x = 20$$

The percent discount on the stereo is 20%.

   b. The amount of tax is the sale price subtracted from the final price, or $203.52 – $192 = $11.52. The tax is computed on the sale price. Write and solve a proportion.

$$\frac{\$11.52}{\$192} = \frac{x}{100}$$

$$11.52 \times 100 = 192 \times x$$

$$1,152 = 192 \times x$$

$$1,152 \div 192 = x$$

$$x = 6$$

The tax rate is 6%.

6. a. First, find the number of fat calories that are in each food item by multiplying the number of fat grams by 9. Then find the ratio $\dfrac{\text{fat calories}}{\text{total calories}}$ for each item.

Hamburger: 9 grams of fat times 9 calories per gram =
81 fat calories

$\dfrac{81 \text{ fat calories}}{260 \text{ calories}} \approx 0.31$, or 31% fat

Cheeseburger: 13 grams of fat times 9 calories per gram =
117 fat calories

$\dfrac{117 \text{ fat calories}}{320 \text{ calories}} \approx 0.37$, or 37% fat

Big Mac: 31 grams of fat times 9 calories per gram =
279 fat calories

$\dfrac{279 \text{ fat calories}}{560 \text{ calories}} \approx 0.50$, or 50% fat

b. From highest to lowest percent of fat, the items are Big Mac (50% fat), cheeseburger (37% fat), and hamburger (31% fat).

7. a. There were 120 boys in the Little League in 1998. There were a total of 120 boys + 80 girls = 200 children in the Little League in 1998. The ratio is $\dfrac{120}{200}$.

$$120 \div 200 = 0.6 = 6\%$$

6% of the children in the Little League in 1998 were boys.

b. The increase in the number of girls in the Little League from 1998 to 2000 was 120 – 80 = 40 girls. The ratio is $\dfrac{40}{80}$.

$$40 \div 80 = 0.50 = 50\%$$

There was a 50% increase in the number of girls in the Little League from 1998 to 2000.

8. a. First, find the total amount of money in Caroline's weekly budget. Add all of the amounts: $200 + $120 + $40 + $80 + $100 + $50 + $64 + $146 = $800. For utilities, the ratio is $\dfrac{120}{800} = 0.15$, or 15%.

15% of Caroline's budget is spent on utilities.

b. Find 8% of $800.

$$0.08 \times \$800 = \$64$$

Caroline budgets $64, or 8% of her total weekly budget for her school loan.

c. Find 25% of 800.

$$0.25 \times \$800 = \$200$$

The proportion of the graph representing rent stands for 25% of Caroline's budget.

Set # 15, page 161

You may use other plans than those that are used here. That is fine.

1. The total number of voters is known (200) and the percent who voted for Tara is known (48%). To find the number of people (part) who voted for Tara, use the formula from Plan Two.

   Part = Percent × Whole = 48% × 200 = 0.48 × 200 = 96

   96 of the voters voted for Tara.

   CROSS OUT 96 IN THE SOLUTION BOX.

2. The part of the cost that Carmine has saved for the trip is known ($630). The percent of the total cost that she has saved is also known (75%). The total or whole is unknown. Write and solve a proportion. Let t stand for the total cost.

$$\frac{630}{t} = \frac{75}{100}$$

$$630 \times 100 = 75 \times t$$

$$63,000 = 75 \times t$$

$$63,000 \div 75 = t$$

$$t = \$804$$

Carmine needs $804 – $630 = $174 more for the trip.

CROSS OUT 174 IN THE SOLUTION BOX.

3. The part of the fifth grade students going to East Middle School is known (144). The percent of the students going to East Middle School is also known (45%). The total number of students in the fifth grade is not known. Write and solve a proportion. Let t stand for the total number of fifth grade students.

$$\frac{144}{t} = \frac{45}{100}$$

$$144 \times 100 = 45 \times t$$

$$14{,}400 = 45 \times t$$

$$14{,}400 \div 45 = t$$

$$t = 320 \text{ students}$$

There is a total of 320 students in the fifth grade.

CROSS OUT 320 IN THE SOLUTION BOX.

4. Change 20% to its fractional equivalent of $\frac{1}{5}$. Find $\frac{1}{5}$ of 84: $84 \div 5 = 16.80$. Subtract the savings of $16.80 from the regular price of $84 to get the sale price of $67.20.

CROSS OUT 67.20 IN THE SOLUTION BOX.

5. For Week 1, if you are saving 10% then you are paying 90% of the price.

90% of $160 = $0.9 \times \$160 = \$144.00$ (Week 1 price)

For Week 2, you are saving 15% more, so you are paying 85% of the price of Week 1.

85% of $144 = $0.85 \times \$144 = \$122.40$ (Week 2 price)

For Week 3, saving 20% more is paying 80% of the Week 2 price.

80% of $122.40 = $0.8 \times 122.4 = \$97.92$ (Week 3 price)

The price of a $160 suit at Shopper's Paradise in Week 3 is $97.92.

CROSS OUT 97.92 IN THE SOLUTION BOX.

6. Use the equation 150% × last year's number on the team = 45 (the number of students on this year's team). Let n stand for last year's number.

$$150\% \times n = 45$$

Change 150% to a decimal.

$$1.50 \times n = 45$$
$$n = 45 \div 1.5 = 30$$

There were 30 students on last year's team.

CROSS OUT 30 IN THE SOLUTION BOX.

7. 10% of a $24.00 bill is $2.40. (Remember, move the decimal point one place to the left.)
5% of a $24.00 bill is $1.20. (Divide the 10% amount by 2.)
15% of a $24.00 bill is $2.40 + $1.20 = $3.60.
A 15% tip on the dinner bill at Sid's Spaghetti Spot is $3.60.

CROSS OUT 3.60 IN THE SOLUTION BOX.

8. Use the fractional equivalents for each percent.

$33\frac{1}{3}\% = \frac{1}{3}$; $\frac{1}{3}$ of 42 = 42 ÷ 3 = 14 tickets sold by Holden

$25\% = \frac{1}{4}$; $\frac{1}{4}$ of 60 = 60 ÷ 4 = 15 tickets sold by Lily

$12\frac{1}{2}\% = \frac{1}{8}$; $\frac{1}{8}$ of 72 = 72 ÷ 8 = 9 tickets sold by Lucinda

The person who sold the most tickets, Lily, sold 15 tickets.

CROSS OUT 15 IN THE SOLUTION BOX.

9. Find the amount the students would like to sell over the next 3 months.
April: Add 10% of $50 to the March amount of $50.

$$0.1 \times \$50 = \$5; \$5 + \$50 = \$55$$

May: Add 10% of $55 to April's amount of $55.

$$0.1 \times \$550 = \$5.50; \$5.50 + \$55 = \$60.50$$

June: Add 10% of $60.50 to May's amount of $60.50.

$$0.1 \times \$60.50 = \$6.05; \$6.05 + \$60.50 = \$66.55$$

The students need to sell $66.55 worth of baked goods in June to reach their goal.

CROSS OUT 66.55 IN THE SOLUTION BOX.

10. First find the total amount of money allocated for the baseball budget. The clue is to look at the coach's salary of $3,000, which is $37\frac{1}{2}$% of the total budget. Use a proportion. Let t stand for the total budget.

$$\frac{\$3,000}{t} = \frac{37.5}{100}$$

$$3000 \times 100 = 37.5 \times t$$

$$300,000 = 37.5 \times t$$

$$300,000 \div 37.5 = t$$

$$t = 8,000$$

The total budget is $8,000.
To find the amount allocated for uniforms, find 25% of $8,000.

$$0.25 \times \$8,000 = \$2,000$$

The amount allocated for uniforms is $2,000.

CROSS OUT 2,000 IN THE SOLUTION BOX.

11. Transportation is $12\frac{1}{2}$%, or $\frac{1}{8}$ of the budget.

$$\frac{1}{8} \times \$8,000 = 8,000 \div 8 = \$1,000$$

The amount allocated for transportation is $1,000.

CROSS OUT 1,000 IN THE SOLUTION BOX.

PERKY PERCENTS

12. Use the proportion amount of $\frac{\text{increase}}{\text{original price}} = \frac{\%}{100}$. The amount of increase is not known. Represent with the letter i.

$$\frac{i}{0.96} = \frac{125}{100}$$

$$100 \times i = 0.96 \times 125$$

$$100 \times i = 120$$

$$i = 120 \div 100$$

$$i = 1.2, \text{ or an increase of } \$1.20$$

Add the increase in price to the price two years ago to find the current price.

$$\$0.96 + \$1.20 = \$2.16$$

The current price of gasoline is $2.16.

CROSS OUT 2.16 IN THE SOLUTION BOX.

The only number not crossed out in the solution box is 70.

The human body is 70% water.

Set # 16, page 168

1. The principal is $2,000, the simple interest rate is 6% (0.06), and the time is 1 year. Use the simple interest formula.

$$I = P \times R \times T$$

$$I = \$2,000 \times 0.06 \times 1 = \$120.00 \text{ in interest}$$

Add the interest to the principal.

$$\$120 + \$2,000 = \$2,120$$

After one year, the amount in Mariah's account was $2,120.

2. The principal is \$600, the simple interest rate is 4% (0.06), and the time is 3 months or $\frac{1}{4}$ of a year. Change $\frac{1}{4}$ to the decimal 0.25 for easier calculations. Use the simple interest formula.

$$I = P \times R \times T$$

$$I = \$600 \times 0.04 \times 0.25 = \$6.00 \text{ in interest}$$

Add the interest to the principal.

$$\$6.00 + \$600.00 = \$606.00$$

Dorian withdrew \$606.00 at the end of the three months.

3. Find the interest for the time from June 1 to July 1, one month. The principal is \$500, the yearly interest rate is 5%, and time, though for one month, is based on one year. Use the simple interest formula.

$$I = P \times R \times T$$

$$I = \$500 \times 0.05 \times 1 \text{ year} = \$25.00 \text{ in interest}$$

Divide by 12 to determine the interest for one month.

$$\$25 \text{ a year} \div 12 \text{ months per year} \approx \$2.08$$

Add the \$2.08 to the principal of \$500 to get the new principal of \$502.08.
Repeat the steps for the time from July 1 to August 1.

$$I = P \times R \times T$$

$$I = \$502.08 \times 0.05 \times 1 \approx \$25.10 \text{ in interest}$$

Divide by 12 to determine the interest for one month.

$$\$25.10 \text{ a year} \div 12 \text{ months in a year} \approx \$2.09$$

Add the \$2.09 to the new principal of \$502.08 to get the amount of \$504.17. By August 1, Lyle had \$504.17 in his account.

4. In this problem, the principal (P), the time (T), and the amount of interest (I) are known ($P = \$1,000$; $T = 1$ year; $I = \$55$). The simple annual interest rate (R) is to be found.

PERKY PERCENTS

You can still use the formula of $I = P \times R \times T$, but now you need to solve for the rate.

$$I = P \times R \times T$$

$$\$55 = \$1{,}000 \times R \times 1$$

$$\$55 = \$1{,}000 \times R$$

Divide both sides of the equation by 1,000.

$$R = 55 \div 1{,}000 = 0.055 = 5.5\% = 5\frac{1}{2}\%$$

Kelly's bank offered a simple interest rate of $5\frac{1}{2}\%$ per year.

5. In this problem, the principal (P), the amount of interest (I), and the annual simple interest rate (R) are known ($P = \$1{,}500$; $I = \$45$; $R = 6\%$). The time (T) is to be found. You can still use the formula of $I = P \times R \times T$, but now you need to solve for the time. Remember, the time in the formula is in years.

$$I = P \times R \times T$$

$$\$45 = \$1{,}500 \times 0.06 \times T$$

$$\$45 = \$90 \times T$$

Divide both sides of the equation by 90.

$$T = 45 \div 90 = 0.5 = \frac{1}{2} \text{ year} = 6 \text{ months}$$

Penelope invested the money for 6 months.

CHAPTER SIX

Stupendous Statistics and Prime Probability

Numbers, numbers, numbers
Floating all around;
In sets, tables, and graphs
Statistics can be found.

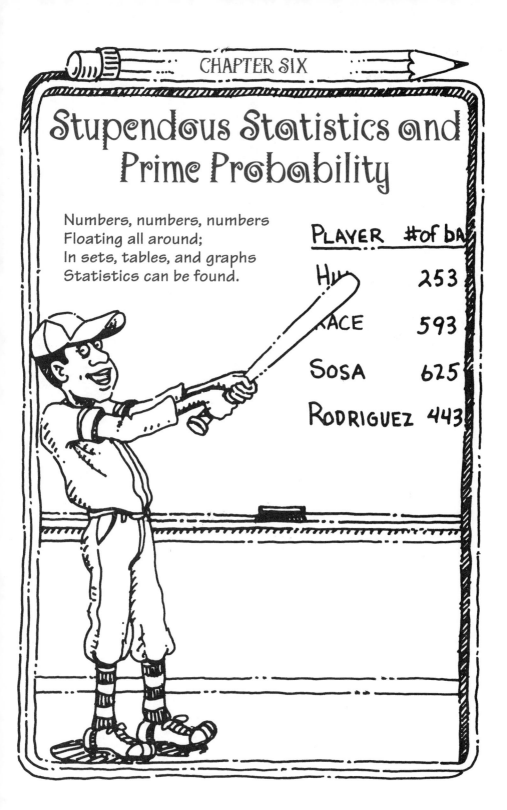

PLAYER	#of ba
HI...	253
...ACE	593
SOSA	625
RODRIGUEZ	443

WORD PROBLEMS WITH STATISTICS

In previous chapters, you have learned to use Polya's Four Steps of Problem Solving. In this chapter, you will supply the missing steps. In the examples, only the solution of the problem will be presented. It is up to you to make sure you read and understand the problem, plan a strategy, and do your plan. You may use a different plan than the one used in the examples. That is fine. Most problems in mathematics have multiple solution methods. That is why it is important that you check your answer.

Statistics is the area of math that deals with collecting and interpreting data. Statistics can appear in tables, graphs, and charts to help describe relationships between and among people, places, and things. Athletes, students, business people, and others use statistics to compare and contrast data about themselves and others.

EXAMPLE:
Use the data in the table below to find the following:
a. Hill's batting average
b. Sosa's number of hits in 1999
c. the number of hits Rodriguez would have needed to achieve a higher batting average than Grace.

1999 Year-End Statistics for the Chicago Cubs Baseball Team

Player	Number of Times at Bat	Number of Hits	Batting Average
Hill	253	76	?
Grace	593	183	?
Sosa	625	?	.288
Rodriguez	443	136	?

The *batting average* of a baseball player is the ratio of the number of hits for the season divided by the number of times at bat.
a. To find the batting average for Hill, find Hill's data in the chart. Form the ratio of hits to times at bat. Round to the nearest thousandth.

$$\frac{76}{253} \approx .300$$

Hill's batting average for 1999 was .300.

b. To find Sosa's number of hits in 1999, find Sosa's times at bat and his batting average in the chart.

Form a proportion with $\dfrac{\text{hits}}{\text{times at bat}} = \dfrac{\text{batting average}}{1}$. Let h stand for the number of hits.

$$\frac{h}{625} = \frac{.288}{1}$$

$$h \times 1 = 625 \times .288$$

$$h = 180$$

Sosa had 180 hits in 1999.

c. To find how many more hits Rodriguez would have needed to have achieved a higher batting average than Grace, first find Rodriguez's batting average and Grace's batting average.

Rodriguez had a batting average of $\dfrac{136}{443} = .307$. Grace had a batting average of $\dfrac{183}{593} = .309$. Form the ratio of hits (h) out of 443 times at bat for Rodriguez: $\dfrac{h}{443}$.

Guess and check values for h (greater than 136) until the ratio equals an average greater than .309 (Grace's batting average).

Try 137: $\dfrac{137}{443} = .309$. That's equal to Grace's batting average.

Try 138: $\dfrac{138}{443} = .312$. That's it!

If Rodriguez had made two more hits, for a total of 138 hits, his batting average would have been greater than Grace's.

EXAMPLE:

The Salty Snack Company took a survey of 120 people, randomly chosen, in order to estimate the average percentage of people who favor salty snacks.

Use the survey results in the table and its corresponding circle graph below to answer these questions:

a. Which snack food belongs in Part B of the circle graph?
b. How much greater was the percentage of those who favored salty snacks than those who favored other snacks?

Survey Results

Snack	Number of People
Potato Chips	40
Cookies	12
Fruit	15
Pretzels	30
Ice Cream	23
Total Surveyed	120

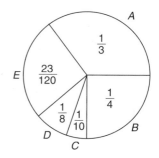

a. Part B is $\frac{1}{4}$ of the graph. To find which snack food belongs in Part B of the circle graph, form a proportion. Let s represent the number of people who favored the snack food in B.

$$\frac{s}{120} = \frac{1}{4}$$

Cross multiply and solve for s.

$$s \times 4 = 120 \times 1$$

$$s = 120 \div 4 = 30 \text{ people}$$

The snack that is favored by 30 people is pretzels.
Pretzels belong in Part B.

b. First find the number of people who favored salty snacks.
Pretzels and potato chips are salty snacks.

$$\text{Pretzels} + \text{chips} = 30 + 40 = 70$$

Now find the percentage of the people who favored salty snacks.

$$\frac{70}{120} \approx 58\%$$

The other snacks are cookies, fruit, and ice cream.
Find the total number and the percentage of people who favored these other snacks.

$$\text{Cookies} + \text{fruit} + \text{ice cream} = 12 + 15 + 23 = 50$$

$$\frac{50}{120} \approx 42\%$$

Subtract to find the percentage more who favored salty snacks than other snacks.

$$58\% - 42\% = 16\%$$

16% more people favored salty snacks than other snacks.

BRAIN TICKLERS
Set # 17

Solve each word problem using statistics. Read the problem carefully and thoroughly, plan a strategy, do the plan, and check your work.

1. Use the data about tall buildings in the table below to answer these questions:

   a. How much taller is one story in the Sears Tower than one story in One World Trade Center? (Round your calculations to the nearest tenth of a foot.)

   b. Rank the buildings by the height of their stories.

Tall Buildings in the United States

Building	Location	Height (in feet)	Number of Stories
Sears Tower	Chicago, IL	1,450	110
Amoco	Chicago, IL	1,136	80
One World Trade Center	New York, NY	1,368	110
Empire State Building	New York, NY	1,250	102

2. Use the census data in the table below to answer these questions:
   a. How many more people lived in Houston in 1998 than in Columbus and Boston combined?
   b. Which city has the greatest percentage of TV stations to total communications stations (TV and radio)?

July 1998 Census Statistics

City	Population	TV Stations	Radio Stations
Houston, TX	1,786,691	15	54
Columbus, OH	670,234	8	25
Boston, MA	555,447	12	21

3. Use the information in the scatter plot below to answer these questions:
   a. How old was the runner who won the race?
   b. What percentage of all of the runners were older than 20 years old?
   c. How long did it take the slowest 25-year-old to finish the race?

Road Race Statistics

(scatter plot: Time (minutes) vs. Age (years))

4. Use the baseball statistics in the table below to answer these questions:
   a. What percent of the Yankees' games, to the nearest percent, were won at home?

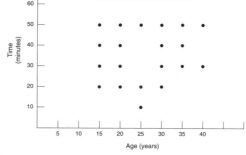

b. What percent of their total games, to the nearest percent, did the Blue Jays win?

c. Rank the teams by their percentage of wins.

1999 American League Baseball Statistics

Team	Wins	Losses	Wins at Home
Toronto Blue Jays	84	78	40
New York Yankees	98	64	48
Boston Red Sox	94	68	49

(Answers are on page 198.)

Word problems with statistics
Can be a painless road;
Let's see how easily we can find
Mean, median, range, and mode.

THE THREE FABULOUS AVERAGES

Sets of numbers are often described and compared by means of *averages*, also called *measures of central tendency*. The most commonly used measures of central tendency are the mean, the median, and the mode.

MATH NOTE

When the word *average* is used with a statistic, it may be unclear which measure of central tendency was used. There are advantages and disadvantages for each measure. Think about this as you learn more about each measure.

Finding the mean of a set of numbers

The most commonly used average is the *arithmetic mean*, also simply called the *mean*. To determine the mean, add up all of the numbers in the data, and then divide the sum by the number of numbers. For example, to find the mean of the five numbers 35, 42, 38, 49, 36, add the numbers, then divide the sum by five.

$$35 + 42 + 38 + 49 + 36 = 200$$

$$200 \div 5 = 40$$

The mean of the five numbers is 40.

Finding the median of a set of numbers

The *median* is the value of the middle number when the numbers are numerically ordered from least to greatest, or greatest to least. The first step in finding the median is to order the numbers. For example, to find the median of the numbers 35, 42, 38, 49, and 36, first order the numbers from least to greatest.

$$35 \quad 36 \quad 38 \quad 42 \quad 48$$

(You can also order the numbers from greatest to least and get the same answer.) There is an odd number of numbers in the set. Therefore, there is a number in the middle. For this set, the middle number is 38, with two numbers less than 38 and two numbers greater than 38. The median of the numbers 35, 42, 38, 49, and 36 is 38.

When you have the same number of numbers on either side of the middle number, you will have found the median. There will always be a middle number when the number of numbers is odd.

What shall we do when the number of numbers is even?

Let us look at an example. To find the median of the numbers 59, 67, 81, 61, again order the numbers from least to greatest.

$$59 \quad 61 \quad 67 \quad 81$$

Notice that there is no middle number. If you draw a line in the middle, the line will be between 61 and 67, with two numbers to the left and two numbers to the right of the line.

$$59 \quad 61 \mid 67 \quad 81$$

The median is the mean of the two numbers on either side of the middle line.

$$(61 + 67) \div 2 = 128 \div 2 = 64$$

The median of the numbers 59, 67, 81, and 61 is 64.

MATH NOTE

When there is an even number of terms in a list, there will
not be one exact middle term, but two terms on either
side of the middle. Find the mean of these two numbers to
find the median. The median may or may not be one of
the numbers in the list, but it will be equal or close to the
middle numbers.

Finding the mode of a set of numbers

The mode is the number, or item, that appears the most times in
a set of data. In a set of data there may be no, one, or more than
one mode. These cases are illustrated in the sets of data below.

The set of numbers 11, 14, 15, 17, 14, 21, 9, 21 has two modes,
14 and 21. Each of these numbers occurs twice in the data,
whereas the other numbers occur only once. The set of numbers
82, 76, 73, 69, 86, 71 has no mode. No one piece of data occurs
more than once. The set of numbers 65, 72, 65, 75, 78, 65, and 75
has one mode. The number 65 occurs three times, more than any
other number in the set.

THE RANGE: A MEASURE OF THE SPREAD OF THE DATA

Knowing the average of a set of data is important. However,
many sets of data can have the same mean. For example, the two
rows of data below have the same mean, 85. (Check this!)

| 77 | 81 | 85 | 89 | 93 |
| 60 | 75 | 87 | 99 | 104 |

The two rows of data differ in the amount of *spread*.

The *range* of a set of numbers is a measure of the spread of
the data. The range is the difference between the highest and
lowest numbers in the set. For example, the range of the first
row of data above is the highest score minus the lowest score, or
93 − 77 = 16. The range of the second row of data is 104 − 60 = 44.
Therefore, reporting both the mean and the range of a set of data

gives a better summary of the data than just reporting the mean alone. The greater the range, the greater the spread in the data.

> Let's see how easy it can be,
> To use statistics painlessly!

EXAMPLE:

Percy Primer scored 87, 85, 87, 84, 95, 89, and 61 on his math quizzes.
a. Find the mean of Percy's scores.
b. Find the median of Percy's scores.
c. Find the mode of Percy's scores.
d. Find the range of Percy's scores.

a. To find the mean, add up Percy's scores on his math quizzes.

$$87 + 85 + 87 + 84 + 95 + 89 + 61 = 588$$

There are seven math scores. Divide the sum by seven.

$$588 \div 7 = 84$$

The mean of Percy's math scores is 84.

b. To find the median of Percy's scores, order the scores from least to greatest.

$$61 \quad 84 \quad 85 \quad 87 \quad 87 \quad 89 \quad 95$$

Find the middle score in the ordered list. The middle score is 87, with three scores to the left of 87 and three scores to the right of 87.
The median of Percy's math scores is 87.
Notice that the mean of Percy's scores (84) is less than the median (87). Let's see why such a large difference occured. Percy has one score of 61, which is much lower than all of his other scores. Find the mean not counting the 61.

$$(87 + 85 + 87 + 84 + 95 + 89) \div 6 = 527 \div 6 \approx 88$$

Without the 61, Percy's mean would be 88, which is close to his median of 87. When one number in a set is much lower or higher than all of the other numbers, the median, and not the mean, may be the better average indicator to use.

c. To find the mode of Percy's scores, survey the list for common numbers.

87   85   87   84   95   89   61

The number 87 appears twice, which is more times than any other number.
The mode is 87.

d. To find the range of Percy's scores, subtract the least number from the greatest number.

87   85   87   84   95   89   61

The least number is 61 and the greatest number is 95.

$$95 - 61 = 34$$

The range of Percy's scores is 34.

In tests, sports, money, and more,
Calculating statistics need not be a chore.

BRAIN TICKLERS
Set # 18

Solve each word problem using the mean, median, mode, or range. Read the problem carefully and thoroughly, plan a strategy, do the plan, and check your work.

1. In the last six seasons of baseball, Frank Foul hit 14, 21, 18, 23, 15, and 29 doubles. Find the mean number of doubles that Frank hit per season.

2. In the last seven basketball games, Sheryl Shot scored 30, 21, 29, 28, 10, 27, and 23 points. How much greater is Sheryl's median than her mean?

3. Use the bake sale price chart below to answer these questions:
   a. What is the range of prices at the bake sale?
   b. What is the mean price?
   c. What must be the cost of a slice of banana bread for the mean of all five prices to be $0.55?

Bigelow Middle School Bake Sale			
Brownies	$0.55	Cookies	$0.35
Cupcakes	$0.60	Drinks	$0.50

4. Use the school transportation data in the tally chart below to answer these questions:
   a. How many students were surveyed?
   b. Which form of transportation represents the mode of the data?
   c. Use the mode to make a conclusion about how most students get to school.

Class Survey on Transportation to School

Form of Transportation	Number of Students	
Bus	⊦⊦⊦⊦		
Walk	⊦⊦⊦⊦ ⊦⊦⊦⊦	
Bicycle			
Driven	⊦⊦⊦⊦ ⊦⊦⊦⊦		

5. In a bowling tournament of four games, Polly Pinly has scores of 95, 110, and 101 on her first three games. What is the minimum score that Polly can achieve in the last game for her mean to be at least 105?

6. The heights of six friends are 5 feet 2 inches, 4 feet 7 inches, 5 feet 2 inches, 4 feet 9 inches, 5 feet 1 inch, and 4 feet 11 inches. What is the median height of the friends?

7. Use the test scores in the table below to answer the following question.
   In calculating final grade averages, a high school science teacher counts quizzes one time, tests two times, and the final test three times. Who has the highest average (mean), Venus or Selena?

Student	Quizzes	Tests	Final Test
Venus	88, 84, 79	90, 87	93
Selena	90, 92, 91	91, 82	85

8. Photo-To-Go offers a cash bonus to its best employee. From the data in the table below, who would you pick as Photo-To-Go's best employee? (Use the mean and/or median to help you decide.)

Employee	Cameras Sold per Month					
	Jan.	Feb.	March	April	May	June
Fiona Film	50	65	23	18	21	50
Fannie Flash	36	40	36	39	42	45

9. Allyn Atlas has history test scores of 83, 91, 81, 91, and 90. He has to take one more history test and wants to end up with an A average (an average of 90–93). Is it possible for him to achieve an A average? If so, what would he have to score on his last history test?

(Answers are on page 200.)

WHAT ARE THE CHANCES? PROBABILITY AT A GLANCE

Take a chance and you will see,
How statistics are used in probability.

When you know how likely it is that something could happen, you are working with *probability*. Probability is used in sports, weather reports, games, surveys, and more. The probability that some particular outcome (called an *event*) will occur is expressed as a ratio.

Hints for making a probability ratio

The numerator of a probability ratio is the number of ways that the event of interest can occur. The denominator is the total number of outcomes that are possible. So, the probability of an event happening is given by this ratio:

$$\frac{\text{number of ways that the event can happen}}{\text{number of outcomes possible}}$$

For example, suppose that you toss a die. What is the probability that you will get a number greater than four?

The die can fall with a one, two, three, four, five, or six on the top face. All of these ways have the same chance of occurring. They are *equally likely*. The total number of equally likely ways the die can fall is six. The outcomes greater than four are five and six. These outcomes are equally likely. There are two equally likely outcomes greater than four. The probability of getting a number greater than four is $\frac{2}{6} = \frac{1}{3}$.

MATH NOTE

> When using the probability ratio, you must make sure that the different ways that the event can occur are equally likely to occur. Also, all the possible outcomes must be equally likely.

EXAMPLE:

What is the probability of spinning a 2 in one spin of the wheel?

First decide on all of the possible numbers on which the spinner can stop (eight possibilities). Decide if each number has the same chance of being spun. (Yes! The spinner is made of eight equal parts.) Of these eight ways, there is only one way the spinner can stop on 2.

Form the probability ratio. The probability of spinning a 2 in one spin is $\frac{1}{8}$.

MATH NOTE

Sometimes a probability is expressed as a fraction and sometimes as a percent. In the preceding example, $\frac{1}{8} = 0.125 = 12\frac{1}{2}\%$ chance of spinning a 2.

EXAMPLE:

Use the data in the table below to answer the following probability question.

If only one member of the chorus can sing in the State Chorus, what is the probability of a sixth grader being picked, at random, to sing in the State Chorus?

Chorus Members		
Grade 6	Grade 7	Grade 8
15	20	10

First find the number of sixth graders (15).
Then find the total number of students, only one of whom will be picked.

$$15 + 20 + 10 = 45$$

The probability ratio is $\frac{15}{45} = \frac{1}{3}$.

There is a $\frac{1}{3}$, or one out of three, chance that a sixth grader will be picked for the chorus. You can also say that there is a $33\frac{1}{3}\%$ chance.

EXAMPLE:

If a dime and a nickel are tossed, what is the probability that at least one coin will land heads up?

First find the different ways the coins could
land. Make a list of all of the possible ways.

Dime	Nickel
H	H
H	T
T	H
T	T

There are four different, equally likely ways the coins could land.

Count the tosses that contain at least one tail (3). The probability
ratio is $\frac{3}{4}$, or 3 out of 4.

Now consider this: Is it possible to toss two coins four times and
have them land H H (no tails) each of the four times? Yes, it is
possible. It could happen even though the mathematical probability is very small. Probability tells us what will happen *in the
long run*. A probability of 3 out of 4 means that if the two coins
were tossed many, many times, about 75% of the total outcomes
would have at least one tail.

MATH NOTE

The probability of an event is a ratio between 0 and 1, or
a percent between 0% and 100%.
If an event will definitely happen, its probability is 1, or
100%. If an event can never happen (impossible!), its
probability is 0, or 0%.

Relative frequency

There are some situations where we know the possible outcomes
even without performing the action. Examples of these situations
are tossing a coin or die, or drawing a card from a well-shuffled
deck of cards. At other times, we must use past information to
determine the likelihood or probability of something happening.
In this case, the probability ratio is called the *relative frequency*.
The relative frequency is a way to predict the likelihood or
probability of something happening using how many times the
event has occurred in the past.

EXAMPLE:

Last year, Billy Ballman got hits in 60 of the 150 games in which he played. If he plays in 160 games next year, estimate the probability of Billy getting a hit in the first game. How many hits should he expect to get in all of the games?

First find the total number of games Billy played in last year (150). Then find in how many games he got a hit (60).
Form a relative frequency ratio of times Billy got a hit last year to total games played last year.

$$\frac{60}{150} = \frac{2}{5} = 0.4 = 40\%$$

There is a 40% chance that Billy will get a hit in the first game. If Billy plays 160 games next year, he should get about 40% of 160 $= 0.4 \times 160 = 64$ hits.

MATH NOTE

Remember that relative frequency is an estimate based on previously collected data.

BRAIN TICKLERS
Set # 19

Solve each word problem by planning and carrying out a strategy. Remember to check your work. At the end of each problem you will find a letter. By placing this letter on the line above the corresponding answer in the answer code at the end of the problem set, you will be able to answer this question:

Which animal is the world's largest mammal?

1. Use the spinner at the right to answer this question: What is the probability of landing on a red space in one spin of the wheel? (L)

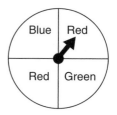

2. What is the probability of randomly picking a face card from a standard deck of 52 cards? (E)

3. In a game made of 26 cards, each with a different letter of the alphabet, the player to first pick all of the different letters that are contained in his or her state wins the game. What is the probability that a player from Indiana will draw one of the letters in "Indiana" in the game's first draw of a card? (T)

4. A red die and a green die are tossed. What is the probability that a sum of seven will be tossed? (H)

5.

Survey of Favorite TV Shows

Show	Votes
Captain Speedy	20
NYC Blues	75
Make Me Rich!	105

If 600 people had been surveyed, what is an estimate of how many people would pick *Captain Speedy* as their favorite show? (L)

6. Karenna, Kandy, and Kayla are vying for two tickets to the Korn Kusp concert. If only two of them can go to the concert, what is Kayla's chance of going to the concert (give your result as a fraction)? (E)

7. There are 30 red candies, 20 yellow, 20 purple, 10 green, 10 orange, and 10 blue candies in a bag of candy.
   a. What is the probability of randomly picking a red candy from the bag? (E)
   b. What is the probability of picking a brown candy? (U)
   c. What is the probability of *not* picking a green candy? (A)

8. Last year it rained 150 days in the town of Dryville. What is the best estimate of the chance for rain tomorrow? (H)

9. Andrew Athlete wants to enter three events in his school's Olympics. If the Olympics consist of swimming, diving, track, and archery, what is the likelihood that Andrew will enter the swimming competition if he chooses the three events at random? (B)

10. In order to win the game, LaToya needs to roll at least a 3 with one roll of a die. What are her chances (%) of winning the game? (W)

Answer Code

The world's largest mammal is ___ ___ ___
$\frac{2}{13}$ 41% $\frac{2}{3}$

___ ___ ___ ___ ___ ___ ___ ___ ___.
$\frac{3}{4}$ 60 0 23% $66\frac{2}{3}$% $\frac{1}{6}$ 90% 50% 20%

(Answers are on page 202.)

BRAIN TICKLERS—THE ANSWERS

Set # 17, page 184

1. a. To find the height of one story, divide the building's height by the total number of stories.

   Sears Tower: 1,450 feet ÷ 110 stories ≈ 13.2 feet per story

   One World Trade Center :
   1,368 feet ÷ 110 stories ≈ 12.4 feet per story

   A Sears story is about 13.2 − 12.4 = 0.8 feet taller than a Trade Center story.
   b. Find the height of one story in the other buildings.

   Amoco: 1,136 feet ÷ 80 stories ≈ 14.2 feet per story

   Empire State Building:
   1,250 feet ÷ 102 stories ≈ 12.3 feet per story

   The order is Amoco (14.2 feet), Sears Tower (13.2 feet), Empire State Building (12.4 feet) and One World Trade Center (12.3 feet).

2. a. First find the combined population of Columbus and Boston.

$$670{,}234 + 555{,}447 = 1{,}225{,}681$$

Subtract this combined population from the population of Houston.

$$1{,}786{,}691 - 1{,}225{,}681 = 561{,}010$$

There were 561,010 more people in Houston than in Columbus and Boston combined.

b. Form the ratio of number of TV stations to total number of TV and radio stations for each of the cities.

Houston: 8 to $(8 + 25) = 8$ to $33 = \dfrac{8}{33} \approx 24\%$

Columbus: 15 to $(15 + 54) = 15$ to $69 = \dfrac{15}{69} \approx 22\%$

Boston: 12 to $(12 + 21) = 12$ to $33 = \dfrac{12}{33} \approx 36\%$

Boston has the greatest percentage of TV stations to total communications stations.

3. a. The person with the lowest time wins the race. Look at the y-axis. Look at the lowest dot, which matches up with the lowest time. Find its match on the x-axis. A 25-year-old won the race.

b. Locate the 20 on the x-axis. Count all of the dots in the plot that are to the right of the 20 on the x-axis. Form a ratio:

$$\frac{\text{greater than 20}}{\text{all people (dots)}} = \frac{12}{20} = 0.6$$

So, 60% of the people were older than 20.

c. Locate the age of 25 on the x-axis. Move straight up to the top dot corresponding to a 25-year-old. The top dot represents the person with the greatest time. This 25-year-old ran the slowest. The time was 50 minutes.

4. a. First find the total number of games the Yankees played. The Yankees played $98 + 64 = 162$ total games. The number of games the Yankees won at home was 48. The ratio of

games won at home to games played is $\dfrac{48}{162}$, or about 30%.

b. The ratio of wins to total games for the Blue Jays was

$\dfrac{84}{(84+78)} = \dfrac{84}{162} \approx 0.52$. The Blue Jays won about 52% of their total games played.

c. Find the percentage of wins for the Yankees and the Red Sox. Yankees wins to total games:

$$\dfrac{98}{(98+64)} = \dfrac{98}{162} \approx 60\%$$

Red Sox wins to total games: $\dfrac{94}{(94+68)} = \dfrac{94}{162} \approx 58\%$.

The order from greatest to least percentage of wins is Yankees (60%), Red Sox (58%), and Blue Jays (52%).

Set # 18, page 190

1. Frank's mean is $(14 + 21 + 18 + 23 + 15 + 29) \div 6 = 120 \div 6 = 20$ doubles.

2. First order the point scores.

$$30 \quad 29 \quad 28 \quad 27 \quad 23 \quad 21 \quad 10$$

There are 7 scores in all. The middle score is 27, which is the median.
The mean is $(30 + 29 + 28 + 27 + 23 + 21 + 10) \div 7 = 168 \div 7 = 24$.
The median is $27 - 24 = 3$ points higher than the mean.

3. a. The range is the highest price ($0.60) minus the lowest price ($0.35).
The range is $0.60 – $0.35 = $0.25.

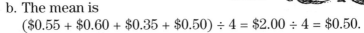

b. The mean is
($0.55 + $0.60 + $0.35 + $0.50) ÷ 4 = $2.00 ÷ 4 = $0.50.

c. In order for the mean to be $0.55, the five prices must add up to $5 \times \$0.55$, or $2.75 ($\dfrac{\$2.75}{5}$ prices = $0.55). The four current prices add up to $2.00. Thus, $2.75 − $2.00 = $0.75. The cost of a slice of banana bread must be $0.75 for the mean to be $0.55.

4. a. Four lines with a diagonal slash represent 5 students. Counting all of the tally marks, there were 30 students in the survey.
   b. The mode of the data is the form of transportation with the most tally marks. Being driven to school represents the mode.
   c. As more students in this survey were driven to school than got to school by other forms of transportation, perhaps more students in the total student population will also be driven to school. This is an educated guess based on statistics.

5. To achieve a mean of 105 Polly will need points from all of the games played to equal 105×4 (total games) = 420 points. This will give her an average of $420 \div 4 = 105$ points. Polly currently has scores that total $95 + 110 + 101 = 306$ points. Polly will need to score at least $420 − 306 = 114$ points in her fourth game to achieve an average (mean) of 105 or more.

6. First order the heights from shortest to tallest.
   4 feet 7 inches   4 feet 9 inches   4 feet 11 inches
   5 feet 1 inch   5 feet 2 inches   5 feet 2 inches
   Since there is an even number of heights, an exact middle number cannot be found. You will need to find the mean of the two middle terms. Change these two heights to inches.

   4 feet 11 inches = 59 inches   5 feet 1 inch = 61 inches

   The two middle heights are 59 inches and 61 inches. The mean of the two middle heights is $(59 + 61) \div 2 = 60$ inches. The median height is 60 inches, or 5 feet.

7. Since quizzes count once, tests count twice, and the final test counts three times, there will be $3 \times 1 = 3$ quiz numbers, $2 \times 2 = 4$ test numbers, and $1 \times 3 = 3$ final test numbers, for a total of 10 numbers to use in calculating the average. The total of Venus's 10 scores is $(88 + 84 + 79) + (90 \times 2) + (87 \times 2) + (93 \times 3) = 884$. Venus's average is $884 \div 10 = 88.4$. Selena's total is $(90 + 92 + 91) + (91 \times 2) + (82 \times 2) + (85 \times 3) = 874$.

Selena's average is 874 ÷ 10 = 87.4. Venus has the higher mean.

8. First order the number of cameras sold from least to greatest for each employee. Then find the median and mean number of cameras sold for each employee.

Fiona: 65   50   50   23   21   18

Fiona's mean is (65 + 50 + 50 + 23 + 21 + 18) ÷ 6 = 227 ÷ 6 ≈ 38. Her median is the mean of the two middle numbers: (50 + 23) ÷ 2 = 73 ÷ 2 = 36.5.

Fannie: 45   42   40   39   36   36

Fannie's mean is (45 + 42 + 40 + 39 + 36 + 36) ÷ 6 = 238 ÷ 6 ≈ 40. Her median is (40 + 39) ÷ 2 = 79 ÷ 2 = 39.5. Fannie's mean and median are both higher than Fiona's. If number of sales is the criterion for determining the best employee, then Fannie would be chosen.

9. Allyn's scores add up to (83 + 91 + 81 + 91 + 90) = 436 points. If six tests will be taken, a minimum total point score of 6 × 90 = 540 points will be needed for an A average. Now 540 − 436 = 104. If 100 is the highest possible point score on a test, it will not be possible for Allyn to achieve an A average.

Set # 19, page 196

1. There are four equal sections on the spinner, two of which are red. So, the red portion makes up $\frac{2}{4} = \frac{1}{2}$ of the spinner. Thus, there is a probability of $\frac{1}{2}$, or 50%, of landing on a red space.

Write an L on the line above 50%.

2. There are four suits in a deck of cards. There is a jack, a queen, and a king in each suit. Therefore there are 4 × 3 = 12 face cards per deck. The probability ratio is $\frac{12}{52}$ ≈ 23%. There

is a 23% chance of picking a face card.

Write an E on the line above 23%.

3. Indiana contains four different letters, I, N, D, and A. The

probability ratio is $\dfrac{4 \text{ letters}}{26 \text{ possible letters}} = \dfrac{4}{26} = \dfrac{2}{13}$.

Write a T on the line above $\dfrac{2}{13}$.

4. There are 6 possible tosses per die: a toss of 1, 2, 3, 4, 5, or 6.
    Now 6 possible tosses times 6 possible tosses equals 36 possi-
    ble outcomes per set of two dice (because the toss of one die
    doesn't affect the toss of the other, you can multiply to find
    the total number of outcomes for two dice). The following
    pairs of outcomes give a sum of 7: 1 and 6; 2 and 5; 3 and 4; 5
    and 2; 4 and 3; or 6 and 1. There are 6 ways to toss a sum of
    seven. The probability ratio is $\dfrac{6}{36} = \dfrac{1}{6}$. There is a $\dfrac{1}{6}$ chance of
    tossing a sum of 7.

    Write an H on the line above $\dfrac{1}{6}$.

5. There were 20 + 75 + 105 = 200 people surveyed. The ratio of

votes for *Captain Speedy* to all votes is the ratio $\dfrac{20}{200} = 10\%$.

If 600 people were surveyed, then 10% × 600 = 60 people is

the best estimate of the number of people who will vote for

Captain Speedy.

Write an L on the line above 60.

6. List the ways that two out of the three girls can be chosen.
    Karenna and Kandy   Karenna and Kayla   Kandy and Kayla
    Of the three ways that two of the girls can be chosen, two of
    the ways include Kayla.

    Thus, there is a $\dfrac{2}{3}$ chance that Kayla will be chosen.

    Write an E on the line above $\dfrac{2}{3}$.

7. a. There is a total of 30 + 20 + 20 + 10 + 10 + 10 = 100 candies
        in the bag. For a red candy, the probability is 20 out of 100,
        or 20%.

Write an E on the line above 20%.

b. Since there are not any brown candies in the bag, the probability of picking a brown candy is 0.
Write a U on the line above 0.

c. There are 10 green candies. Thus 100 – 10 = 90 candies are not green. The probability of not picking a green candy is 90 out of 100, or 90%.
Write an A on the line above 90%.

8. 150 days out of a year of 365 days is $\frac{150}{365} \approx 0.41$. There is about a 41% chance that it will rain tomorrow.
Write an H on the line above 41%.

9. The combinations of events that Andrew could enter are, swim-dive-archery, swim-track-archery, and dive-track-archery. There are four combinations. Three out of the four combinations contain swimming. There is a $\frac{3}{4}$ chance of Andrew choosing swimming.
Write a B on the line above $\frac{3}{4}$.

10. There are 6 possible tosses with one die: 1, 2, 3, 4, 5, or 6.
The numbers 3, 4, 5, and 6 (four numbers) have a value of at least 3. The probability is $\frac{4}{6} = \frac{2}{3}$, or as a percentage, $66\frac{2}{3}\%$.
Write a W on the line above $66\frac{2}{3}\%$.

Answer Code

The world's largest mammal is $\underset{\frac{2}{13}}{\text{T}}$ $\underset{41\%}{\text{H}}$ $\underset{\frac{2}{3}}{\text{E}}$

$\underset{\frac{3}{4}}{\text{B}}$ $\underset{60}{\text{L}}$ $\underset{0}{\text{U}}$ $\underset{23\%}{\text{E}}$ $\underset{66\frac{2}{3}\%}{\text{W}}$ $\underset{\frac{1}{6}}{\text{H}}$ $\underset{90\%}{\text{A}}$ $\underset{50\%}{\text{L}}$ $\underset{20\%}{\text{E}}$.

CHAPTER SEVEN

Jumping Geometry Jubilee

Perimeter, area, measurement, too;
Polygons and circles—
It's geometry for you.

Lines, angles, graphs, all kinds of shapes, and measurement, too, are part of the study of geometry. Meteorologists, landscapers, engineers, architects, and other professionals use geometry. Following are some of the important geometry terms that they use in their jobs.

JUBILEE OF GEOMETRY VOCABULARY

Polygon: a two-dimensional *closed* shape made of at least three line segments, none of which cross any of the other segments

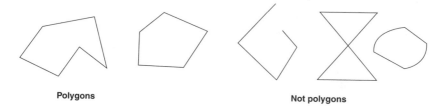

Polygons   Not polygons

It is important to know when a shape is not a polygon, as well as when a shape is a polygon. In your own words, tell why the last three figures above are not polygons.

Regular polygon: a polygon that has sides that are all of equal length (*congruent sides*), and angles that all have the same measure (*congruent angles*)

Congruent figures: figures that can be moved, translated, rotated, or flipped to fit exactly on top of each other

Perimeter: the total distance around a closed figure

Area: a measure of the space enclosed by a two-dimensional shape, measured in *square units*

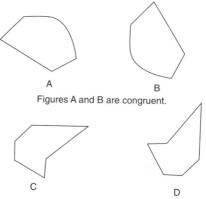

A   B

Figures A and B are congruent.

C   D

Figures C and D are congruent.

Circle: the set of points, in a plane, that are equal in distance from a given point (the *center* of the circle)
The next three terms relate to circles.

Circumference: the distance around a circle

Diameter: a line segment from one point on a circle to another point on the circle that passes through the center

Radius: a line segment from the center of a circle to any point on the circle (half of a diameter of the circle)

More of the geometry jubilee: names of polygons

Some polygons have special names that correspond to the number of sides and angles.

Number of Sides and Angles	Name	Number of Sides and Angles	Name
3	Triangle	7	Heptagon
4	Quadrilateral	8	Octagon
5	Pentagon	9	Nonagon
6	Hexagon	10	Decagon

So many words to learn and use.
If we learn a few more,
through problems we'll cruise!

Special members of the quadrilateral family

Did you notice that squares and rectangles are not in the polygon list above? Squares and rectangles have four sides, and so are members of the quadrilateral family. Following are the definitions of special quadrilaterals.

Parallelogram: a quadrilateral with opposite sides that are parallel and congruent

Sides *AB* and *DC* are parallel and
congruent.
Sides *AD* and *BC* are parallel and
congruent.

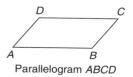

Parallelogram *ABCD*

Rectangle: a parallelogram with four right angles (angles that
measure 90 degrees)
Angles *A*, *B*, *C*, and *D* are all right angles.
Opposite sides are parallel and congruent.

Rectangle *ABCD*

Square: a rectangle with congruent sides
All sides are congruent.
All angles are right angles.
Opposite sides are parallel.

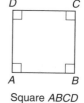

Square *ABCD*

Rhombus: a parallelogram with all four sides congruent
All sides are congruent.
Opposite sides are parallel.

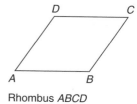

Rhombus *ABCD*

Circles, squares, angles, lines,
Enough of the words!
To solve problems, it's time!

WORD PROBLEMS WITH GEOMETRY

In this chapter, as in Chapter Six, you will supply the missing steps
in Polya's Four Steps of Problem Solving. In the examples, only
the solution of the problem is presented. You must make sure you
read and understand the problem, plan a strategy, and do your
plan. Be sure to check your answer to each of the problems.

EXAMPLE:

Vicki is sewing a rectangular quilt that will measure 12 feet by 8 feet when completed. The quilt will be made from square patches, 4 feet by 4 feet, sewn together.

a. How many square patches 4 feet by 4 feet will Vicki need to make the quilt?

First notice that the length of the quilt is 12 feet. The length of each square patch is 4 feet. There will be $12 \div 4 = 3$ patches going along the length.
The width of the quilt is 8 feet. There will be $8 \div 4 = 2$ patches going along the width.
It is usually wise to draw a picture when geometry or measurement ideas are involved.
Count the square patches. There are 6 of them. (You can also multiply the length by the width, 2×3, for a product of 6 square patches.)

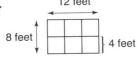

Vicki will need six 4 feet by 4 feet squares to make the quilt.

b. Vicki wants to sew a square design, consisting of four patches, onto her quilt.
Vicki is not sure where to put the two-patch-by-two-patch design.
How many large squares, two patches long and two patches wide, are contained in the quilt?

One way to answer this question is to use a pencil to outline all of the specified shapes. Count as you go.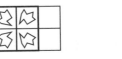
There are two large squares, two patches long and two patches wide, contained in Vicki's quilt.
Another method is to write a letter in each patch.
Now make a list of all of the letters in regions that together form a two-patch-by-two-patch square.
Patches labeled a, b, d, and e form such a square.

Patches
labeled b, c, e, and f, also form such a square.

MATH NOTE

Labeling units and listing possibilities helps you keep separate those pieces that overlap.

EXAMPLE:

A professional basketball court is a rectangle that measures 85 feet by 46 feet. How many feet of black tape would be needed to cover the perimeter of the court? A rectangle has two sides of equal length and two sides of equal width. The perimeter is the distance around the court.

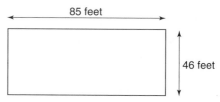

Again, it is wise to draw a picture!
To find the perimeter, you need to add the lengths of all four sides. You can do the arithmetic in different ways. You can add the length of each side.

$$46 + 85 + 46 + 85 = 262 \quad \text{or} \quad 85 + 85 + 46 + 46 = 262$$

You can also find the perimeter using the fact that the rectangle has two sides of length 85 feet and two sides of length 46 feet.

$$(2 \times 85) + (2 \times 46) = 262$$

To cover the perimeter of the court, 262 feet of tape are needed.

MATH NOTE

Drawing a picture is a great plan; you can "see" the problem, making it easier to understand!

Awesome area formulas

Polygon	Formula	
Parallelogram	base × height	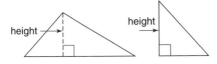
Rectangle	length × width	
Square	side length times side length, or side length *squared*	
Triangle	(length of base × height) ÷ 2, or $\frac{1}{2}$ × (base × height)	

MATH NOTE

The *height* (also called an *altitude*) of a triangle is a line segment drawn from any vertex on the triangle *perpendicular* to the opposite side. (Perpendicular lines meet at right or 90° angles.) Sometimes the height of a triangle lies inside of or on the triangle, and sometimes the height lies outside of the triangle.

Height inside or on a triangle   **Height outside of a triangle**

The units for the area of a two-dimensional shape are often written with an exponent of 2 or by using the word "square." For example, the expressions below are equivalent.

24 square feet   24 sq ft   24 feet^2 24 ft^2

In each case, the expression is read as "24 square feet."

EXAMPLE:

A baseball "diamond" is really a baseball square. A baseball player runs 360 feet around the bases after hitting a home run. What is the area enclosed by a baseball diamond?

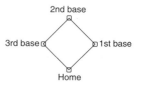

The perimeter of the square is 360 feet. We must find its area. Draw a picture!

Since a square has four congruent sides, the perimeter divided by four is the measure of one side of the square.

$$360 \div 4 = 90 \text{ feet between the bases}$$

Now use the area formula for the area of a square. The area of a baseball square is the side length times the side length.

$$90 \text{ feet} \times 90 \text{ feet} = 8{,}100 \text{ square feet} = 8{,}100 \text{ ft}^2$$

The area of a baseball diamond is $8{,}100 \text{ ft}^2$.

Stupendous circle formulas

Circumference: The circumference of a circle is equal to $\pi \times$ the diameter of the circle, or $2 \times \pi \times$ the radius of the circle. Here is the formula to use:

$$C = \pi \times d = 2 \times \pi \times r$$

Area: The area of a circle is equal to $\pi \times$ the radius squared. Here is the formula to use:

$$A = \pi \times r^2$$

MATH NOTE

What is π? π is a letter of the Greek alphabet, pronounced "pi." In geometry, π represents the ratio of the distance around a circle (the circumference) to the distance across a circle (the diameter). This ratio, π, is constant for all circles, no matter how large or small. The circumference of a circle is a bit more than three times longer than the diameter of the circle. The most commonly used estimate of π is 3.14. π is used to find both the circumference and the area of circles.

EXAMPLE:

A Ferris wheel has a diameter of 50 yards. What distance do you travel in one turn of the wheel? Since the seats are on the outside of a Ferris wheel, the distance you travel in one turn of the wheel is the distance around the wheel, or its circumference. Its diameter is 50 yards.
Draw a picture and include the known information.

50 yd

Use the formula for the circumference of a circle. Use 3.14 for π.

$$C = \pi \times d \approx 3.14 \times 50 \text{ yards} = 157 \text{ yards}$$

The distance you travel in one turn of the Ferris wheel is about 157 yards.

EXAMPLE:

The Westwood High football team wants to create a green felt team banner. The pennant will look like the picture at the right. A circular white patch with a diameter of 8 inches will be sewn onto the green felt. The patch will contain the mascot.

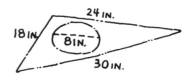

a. What is the area of the white felt patch?
b. How much green felt will be visible on the front of each pennant?

a. Find the area of the white circular patch. The diameter is 8 inches.
The formula for the area of a circle is $A = \pi \times r^2$, where $\pi \approx 3.14$ and r, the radius, is half of the diameter of the circle. In this case, $r = 8 \div 2 = 4$ inches.
The area is about 3.14×4 inches $\times 4$ inches $= 50.24$ square inches.
The area of the white patch is about 50.24 in^2.

b. The area of the green felt that is visible is the area of the green felt triangle minus the area of the white circular patch.

The area of the green felt triangle is $\frac{1}{2} \times$ (base $\times$ height) =
$\frac{1}{2} \times$ (24 inches $\times$ 18 inches) = 216 square inches.

The area of the white patch from Part a is about 50.24 square inches.

The area of the green felt that is visible is about 216 – 50.24 = 165.76 square inches.

Formulas, shapes, and pictures galore,
Practice, practice and through problems you'll soar!

BRAIN TICKLERS
Set # 20

Read each problem. Plan a strategy by using pictures, formulas, or counting parts. Do the plan and check your work.

1. Nina's club has a secret symbol that looks like the picture at the right. The total number of triangles, large and small, in the symbol is the number of members in the club. How many members are in the club?

2. The perimeter of a football field is 317 meters. The length is 109.7 meters. Find the width of the field.

3. In a relay race, each of the four runners on a team must run around a circular track one time. The distance across the track, through the center, is 80 yards. What is the total distance run by each team?

4. A triangular kite has a base of 18 inches and two other sides of 15 inches each. The height of the triangle is 12 inches. How many kites can be made from 500 square inches of fabric?

5. Donna has three pieces of ribbon, each 12 inches in length. She will make borders for three picture frames. The frames are in the shape of a square, a regular hexagon, and an equilateral (regular) triangle. One piece of ribbon is just long enough to go around each frame. What will be the length of ribbon on a side of each frame?

6. The length of the second hand of a clock is six inches. How far does the tip of the second hand travel in one minute?

7. How many $\frac{1}{2}$-inch beads will be needed to make a bracelet with a diameter of four inches?

8. The distance between bases on a Little League field is 60 feet, and on a standard baseball field is 90 feet. A Little Leaguer and a baseball player are both running around their fields. What is the minimum number of times that each will run around his field so that the total distances will be the same?

9. A photograph measuring 3.5 inches by 5 inches will be placed into a wooden rectangular frame measuring 6 inches by 10 inches. What is the area of the wooden part of the frame?

10. A lighthouse flashes a beam all the way around a circle. In the fog, the beam is visible up to 2 miles away. Over what area is the beam visible?

11. Two square emblem designs, each six inches on a side, were submitted in a school contest. The designs will be painted red and blue, as shown.

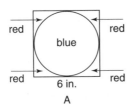

A

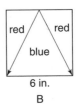

B

a. Which emblem will require the lesser amount of red paint?
b. How much less area will be covered red?

12. An architect is to design a rectangular patio that has a total area of 200 square feet. Each side of the patio will be a whole number of feet. How many different rectangular patios can be designed? (Count, for example, 4 by 50 and 50 by 4 as one patio.)

13. A caterer is setting up four square tables for a banquet. Each table must share one full side with at least one other table. One person will sit at each exposed side of a table. No one will sit at a corner.
a. Find an arrangement of tables that will seat the minimum number of people.
b. Find an arrangement of tables that will seat the maximum number of people.

14. Three squares are placed side by side as shown.
a. If the perimeter of the rectangle is 24 cm, find the area of each square.
b. Find the area of the rectangle.

15. If the sides of a square were each increased by 5 centimeters, its new perimeter would be 36 centimeters. Find the length of a side of the original square.

16. If the sides of a square were each decreased by 3 inches, its new area would be 49 square inches. Find the area of the original square.

(Answers are on page 226.)

Height, weight, money, and time—
Measurement is next: inch, pound, liter, dime.

WORD PROBLEMS WITH MEASUREMENTS

You have already solved word problems with measurement in previous chapters. Let's put it all together, both customary and metric, for a painless pursuit of solving problems.

Common Customary Measurements

Length or Distance	Weight	Capacity
1 foot = 12 inches	1 pound = 16 ounces	1 cup = 8 fluid ounces
1 yard = 3 feet	1 ton = 2,000 pounds	1 pint = 2 cups
1 mile = 5,280 feet		1 quart = 2 pints
		1 gallon = 4 quarts
		1 tablespoon = 3 teaspoons

Marvelous Metric Measurements

Length	Mass
1 meter = 1,000 centimeters	1 kilogram = 1,000 grams
1 kilometer = 1,000 meters	

Curiously Common Conversions

Length or Distance	Mass or Weight	Capacity
1 centimeter = 0.39 inch	1 kilogram = 2.2 pounds	1 gallon = 3.785 liters
1 meter = 39.37 inches		1 liter = 33.8 fluid ounces
1 mile = 1.6 kilometers		
1 inch = 2.54 centimeters		

Temperature

There are two commonly used temperature scales, the Fahrenheit scale and the Celsius scale. Because there is not a direct relationship between the two scales, formulas are needed to convert from one scale to the other.

To change Celsius (°C) to Fahrenheit (°F), use the formula $F = (1.8 \times C) + 32$.

To change Fahrenheit to Celsius, use the formula $C = (F - 32) \div 1.8$.

Ratios and proportions,
With geometry of course,
Are all a part of measurement—
Let's move on with force!

MATH NOTE

You may find it helpful in solving some measurement problems to review Chapter 4 on ratios and proportions.

EXAMPLE:

A man in India once grew a mustache that was 11 feet 11.5 inches long! About how many meters long was his mustache?

Proportions are helpful in changing between customary units and metric units. First you must change 11 feet to inches.

11 feet × 12 inches per foot = 132 inches

Now find the length of the mustache in inches.

132 inches + 11.5 inches = 143.5 inches

Look at the conversion chart to find that one meter equals 39.37 inches.
Use a proportion to change inches to meters. Let m stand for the number of meters.

$$\frac{1 \text{ meter}}{39.37 \text{ inches}} = \frac{m \text{ meters}}{143.5 \text{ inches}}$$

Cross multiply.

$$143.5 = 39.37 \times m$$

Divide both sides of the equation by 39.37.

$$143.5 \div 39.37 \approx 3.64 \text{ meters.}$$

The man's mustache was about 3.64 meters long.

EXAMPLE:

A woman in Massachusetts once did 8,341 somersaults in 10.5 hours. At this rate, how many somersaults could she do in 10 minutes?

Since the problem asks for somersaults per minute, change hours to minutes.

10.5 hours = 10.5 hours × 60 minutes per hour = 630 minutes

Use a proportion to compare somersaults per 630 minutes to somersaults per 10 minutes. Let s stand for the number of somersaults in 10 minutes.

$$\frac{8{,}341 \text{ somersaults}}{630 \text{ minutes}} = \frac{s \text{ somersaults}}{10 \text{ minutes}}$$

Cross multiply.

$$83{,}410 = 630 \times s$$

Divide both sides of the equation by 630.

$$83{,}410 \div 630 \approx 132.4, \text{ or about } 132$$

The woman could do about 132 somersaults in 10 minutes.

EXAMPLE:

Althea is going to visit Frank, stopping at some but not all of her friends' homes along the way. If she always walks toward Frank's house, never back toward her own house, what is the shortest route that she can take?

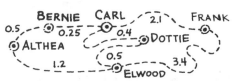

Find all of the different routes that Althea can take to get to Frank's house. One way to do this is to outline all of the routes. Another way is to make a list of all of the routes in a systematic way. Label each friend's home with the first letter of his or her name. The possible routes from A (Althea) to F (Frank) are ABCF, ABCDEF, AEF, and AEDCF.

Calculate the total distance covered in each route.

Route ABCF measures 0.5 mile + 0.25 mile + 2.1 miles = 2.85 miles.

Route ABCDEF measures 0.5 mile + 0.25 mile + 0.4 mile + 0.5 mile + 3.4 miles = 5.05 miles.

Route AEF measures 1.2 miles + 3.4 miles = 4.6 miles.

Route AEDCF measures 1.2 miles + 0.5 mile + 0.4 mile + 2.1 miles = 4.2 miles.

Since 2.85 miles is the shortest distance, route ABCF is the shortest route.

EXAMPLE:

Water boils at a temperature of 100° Celsius. At what temperature Fahrenheit does water boil?

Use the formula for changing temperature on the Celsius scale to temperature on the Fahrenheit scale. (You do not need to memorize formulas or conversion units. You can always refer to the charts in this book or in other references.)

$$F = 1.8 \times C + 32$$

$$F = (1.8 \times 100) + 32$$

$$F = 180 + 32 = 212$$

The Fahrenheit temperature at which water boils is 212°F.

BRAIN TICKLERS
Set # 21

Read each problem. Plan a strategy by using ratios and proportions, or by using formulas. Use the measurement charts to help you. Do the plan and check your work.

1. The tallest man on record was Robert Wadlow. His height was 272 centimeters. About how many feet tall was Robert Wadlow?

2. A dripping faucet can waste about 20 gallons of water per week. About how many liters of water would be wasted in two weeks?

3. The maximum official weight of a golf ball is 1.61 ounces. The maximum official weight of a tennis ball is 2.06 ounces. About how many more golf balls than tennis balls would weigh a total of one pound?

4. The maximum average daily temperature in Tokyo, Japan, in the month of January is 49.1° Fahrenheit. Find this temperature in degrees Celsius.

5. Toenails grow an average rate of 0.004 inches per day. If you didn't cut or break your toenails, how many inches long would they be in one non-leap year?

6. The recipe for Nestle Toll House cookies calls for $2\frac{1}{4}$ cups of flour and yields 5 dozen cookies. If the recipe were increased to yield 160 cookies, how many cups of flour would be needed?

7. Which route from the school to the library will cover a distance of 2.5 miles?

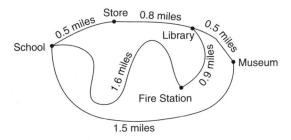

8. Which route represents the shortest distance?

9. The rim of a basketball goal is 10 feet up from the floor of the court. How many meters is the rim from the floor of the court?

(Answers are on page 231.)

One more geometry formula—
Pythagorean is its name;
It's not hard, so get on board
To the path of geometry fame.

TRIANGLES AND THE PYTHAGOREAN THEOREM

The *Pythagorean Theorem* is used to find the length of a side of a right triangle when the lengths of two of the sides are known. The theorem has many uses. It can be used to find the distance between cities on a map, the height of a building from the length of its shadow, the diagonal of a rectangle, and much, much more.

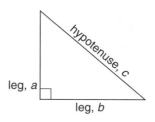

The longest side of a right triangle is called the *hypotenuse* of the triangle. The two shorter sides of a right triangle are called the *legs* of the triangle.

The Pythagorean Theorem states that the sum of the squares on the legs of a right triangle equals the square on the hypotenuse. If a and b represent the length of the legs, and c the length of the hypotenuse, then the Pythagorean Theorem can be stated as follows:

$$a^2 + b^2 = c^2$$

You can "see" the Pythagorean Theorem in the diagram at the right. Squares have been drawn on each of the three sides. The lengths of the two legs are 3 units and 4 units. The length of the hypotenuse is 5 units.

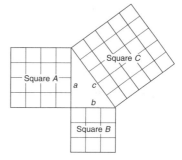

The area of Square A is 4 units × 4 units = 16 square units.

The area of Square B is 3 units × 3 units = 9 square units.

The area of Square C is 5 units × 5 units = 25 square units.

$$a^2 + b^2 = 16 + 9 = 25 = c^2$$

223

EXAMPLE:

Sandy Beach is 6 miles due east of Stone Beach.
Salty Beach is 8 miles due north of Stone Beach.
What is the direct distance from Sandy Beach to
Salty Beach?

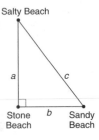

Draw a diagram, as close to scale as possible.
Since due east and due north form right angles,
the distances form the sides of a right triangle.
Side a = 8 miles, side b = 6 miles, and side c is unknown. Use the
Pythagorean Theorem.

$$a^2 + b^2 = c^2$$

$$(8 \times 8) + (6 \times 6) = c^2$$

$$64 + 36 = c^2$$

$$100 = c^2$$

Find the square root of 100. The square root of 100 is 10, since
$10 \times 10 = 100$.
It is 10 miles from Sandy Beach to Salty Beach.

EXAMPLE:

A rectangular garden measures 5 feet by 12 feet. Find the length
of its diagonal.

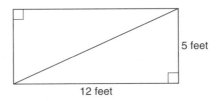

5 feet

12 feet

Draw a picture! The diagonal of a rectangle cuts it into two congruent halves. Each half is a right triangle.
Use the Pythagorean Theorem.

$$a^2 + b^2 = c^2$$

$$(5 \times 5) + (12 \times 12) = c^2$$

$$25 + 144 = c^2$$

$$169 = c^2$$

The square root of 169 is 13, since $13 \times 13 = 169$.
The diagonal is 13 feet long.

MATH NOTE

When using the Pythagorean Theorem, either shorter side may be called side a or side b. However, the hypotenuse must be called side c.

A few more problems
And then we're done.
Learning painless techniques
Can truly be fun!

BRAIN TICKLERS
Set # 22

1. A traffic sign that is eight feet high casts a shadow that measures six feet long. Find the distance from the top of the sign to the end of the shadow.

2. A ladder is tilted so that its top rests 12 feet up a wall. If the bottom of the ladder is five feet from the bottom of the wall, how long is the ladder?

3. A rectangular gift box whose top measures 24 inches by 7 inches will have a piece of ribbon stretching from one corner of the lid to its opposite corner. How many inches long should the ribbon be if it must fit exactly on top of the box with no ribbon left over?

4. Sorange traveled 9 miles south and then headed directly west 12 miles. How far was she, as the crow flies, from her starting point? ("As the crow flies" means the shortest distance between any two points.)

5. From his home, Isaac drove 24 miles east and then 10 miles directly north to reach the concert hall. How far, as the crow flies, is the concert hall from Isaac's home?

(Answers are on page 234.)

BRAIN TICKLERS—THE ANSWERS

Set # 20, page 215

1. You can label each part of the symbol as shown.
   Count the small triangles, each forming one part of the symbol.
   They are pieces *A*, *B*, *C*, *D*, *E*, and *F*, for six triangles.
   Next count the triangles made of two pieces.
   They are *AB*, *DE*, *AD*, and *BE*, for four more triangles.
   Count the triangles made of three pieces.
   They are *ADC* and *BEF*, for two more triangles.
   There are no triangles made of four or five pieces.
   Lastly, count the big triangle, or the symbol itself, for one last triangle. There are 6 + 4 + 2 + 1 = 13 triangles.
   There are 13 members in the club.

2. Draw a picture of a rectangle with a length of 109.7, and *W* as the width.

109.7 meters

W

$$P \text{ (Perimeter)} = L + L + W + W$$
$$317 = 2 \times L + 2 \times W$$
$$317 - (2 \times 109.7) = 2 \times W$$
$$317 - 219.4 = 2 \times W$$
$$97.6 = 2 \times W$$
$$97.6 \div 2 = W$$
$$W = 48.8$$

The width of the rectangle is 48.8 meters.

3. The distance around the track is its circumference (C).

$$C = \pi \times d \approx 3.14 \times 80 \text{ yards} = 251.2 \text{ yards}$$

The total distance run by each team is about 251.2×4 runners per team = 1,004.8 yards.

4. Draw a triangle with a base of 18 inches, two equal-length sides of 15 inches each, and a height of 12 inches. The area of a triangle (A) is $\frac{1}{2} \times$ (base × height).

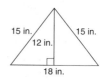

$$A = \frac{1}{2} \times (18 \times 12) = \frac{1}{2} \times 216 = 108 \text{ square inches}$$

500 total square inches ÷ 108 square inches per kite ≈ 4.63 kites

Since you cannot make a part of a kite, four kites can be made.

5. A square has four congruent sides, a regular hexagon has six congruent sides, and an equilateral triangle has three congruent sides.
    With a piece of ribbon 12 inches long, the square is 12 inches ÷ 4 = 3 inches per side.

The hexagon is 12 inches ÷ 6 = 2 inches per side

The triangle is 12 inches ÷ 3 = 4 inches per side

6. The tip of the second hand travels the circumference of a circle in one minute. The radius of the circle is the length of the second hand. The circumference of a circle with a known radius is $2 \times \pi \times r$.

$$C \approx 2 \times 3.14 \times 6 \text{ inches} = 37.68 \text{ inches}$$

In one minute the tip of the second hand travels a distance of about 37.68 inches.

7. The beads form the circumference of the bracelet.

$$C = \pi \times d \approx 3.14 \times 4 \text{ inches} = 12.56 \text{ inches}$$

Since each bead is 0.5 inches long, 12.56 inches ÷ 0.5 inches = 25.12. Since a part of a bead cannot be used, you can either use 25 beads and have a slightly smaller bracelet, or use 26 beads and have a slightly larger bracelet.

8. Running around a ball field means running the distance of the field's perimeter. The perimeter of the Little League field is 60 feet × 4 = 240 feet. The perimeter of the baseball field is 90 feet × 4 = 360 feet. Make a chart to answer the question.

	Little League			Baseball	
Times Around	1	2	3	1	2
Total Distance	240	480	720	360	720

If the Little League player runs around the Little League field three times and the baseball player runs around the baseball field two times, their total distances will be the same.

9. The area of the picture frame equals the combined area of the photo and the wooden frame minus the area of the photo.

Combined area of photo and frame =
6 inches × 10 inches = 60 square inches

Photo area = 3.5 inches × 5 inches = 17.5 square inches

Subtract: 60 square inches – 17.5 square inches = 42.5 square inches.
The area of the wooden part of the frame is 42.5 square inches.

10. The lighthouse is at the center of a circle whose radius is the length over which the beam is visible. The area of a circle is $A = \pi \times r^2$.

$$A \approx 3.14 \times 2 \text{ miles} \times 2 \text{ miles} = 12.56 \text{ square miles}$$

The area covered by the beam is 12.56 square miles.

11. a. One way to find the area covered by red paint is to find the total area of the emblem and subtract the area of the blue part.
Area of emblem A: $A = 6$ inches $\times 6$ inches $= 36$ square inches
Area of the blue (circular) part of emblem A:
$A \approx 3.14 \times 3$ inches $\times 3$ inches $= 28.26$ square inches
Area of the red part of emblem A: $A \approx 36$ square inches – 28.26 square inches $= 7.74$ square inches
Area of emblem B: $A = 6$ inches $\times 6$ inches $= 36$ square inches
Area of the blue (triangular) part of emblem B:
$A = \frac{1}{2} \times (\text{base} \times \text{height}) = \frac{1}{2} \times (6 \text{ inches} \times 6 \text{ inches}) = 18$ square inches
Area of the red part of emblem B: $A = 36$ square inches – 18 square inches $= 18$ square inches
Since 7.74 is less than 18, emblem A will require less red paint than emblem B.
    b. 18 square inches – 7.74 square inches = 25.74 square inches
Emblem A will require 25.74 square inches less of red paint than emblem B.

12. The area of a rectangle is length times width. Make a list to find two whole numbers whose product is 200.

Width	Length
1	200
2	100
4	50
5	40
8	25
10	20

Six possible patios can be designed with an area of 200 square feet.

13. Draw all possible pictures of four attached squares. Each square must share at least one whole side with another square. Put an x on each exposed side.

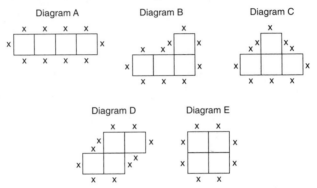

Diagram A   Diagram B   Diagram C

Diagram D   Diagram E

Count the x's. The table arrangements shown in Diagrams A, B, C, and D will each seat ten people. The square arrangement in Diagram E will seat eight people.

a. The minimum number of people that can be seated is eight.

b. The maximum number of people that can be seated is ten.

14. a. Use the picture, and place an x on each exposed side of the three squares. There are eight x's, or eight equal parts that sum to a perimeter of 24 centimeters. Each square must measure 24 centimeters ÷ 8 = 3 centimeters per side. Each square has an area of 3 centimeters × 3 centimeters = 9 square centimeters.

b. The rectangle's length is 3 centimeters + 3 centimeters + 3 centimeters = 9 centimeters. Its width is 3 cm. The area is 9 centimeters × 3 centimeters = 27 square centimeters.

15. Work backwards on this one! The new square has a perimeter of 36 centimeters. Thus each side measures 36 centimeters ÷ 4 = 9 centimeters. Each side was *increased* by 5 centimeters. So *decrease* each new side by 5 to get the side length of the original square.

$$9 \text{ centimeters} - 5 \text{ centimeters} = 4 \text{ centimeters}$$

The length of a side of the original square was 4 centimeters.

16. Work backwards again! A square with an area of 49 square inches has a side length which when multiplied by itself results in an area of 49 square centimeters. As 7×7 equals 49, the new square has a side length of 7 inches.
The new square is smaller per side by 3 inches. Therefore add 3 inches to each side to find the side length of the original square.

$$7 \text{ inches} + 3 \text{ inches} = 10 \text{ inches}$$

The area of the original square is 10 inches × 10 inches = 100 square inches.

Set # 21, page 221

1. Use a proportion to change centimeters to inches.
Look at the conversion chart to find that one inch equals 2.54 centimeters. Let i stand for Mr. Wadlow's height in inches.

$$\frac{1 \text{ inch}}{2.54 \text{ centimeters}} = \frac{i \text{ inches}}{272 \text{ centimeters}}$$

Cross multiply.

$$272 = 2.54 \times i$$

Divide both sides of the equation by 2.54.

$$272 \div 2.54 = i$$

$$i \approx 107.09 \text{ inches}$$

Now change the number of inches to feet.

$$107.09 \text{ inches} \div 12 \text{ inches per foot} \approx 8.9 \text{ feet}$$

Robert Wadlow was about 8.9 feet tall, or almost 9 feet tall.

2. Look at the conversion chart to find that one gallon equals 3.785 liters.
Use a proportion to convert gallons to liters. Let L stand for the number of liters wasted in one week.

$$\frac{1 \text{ gallon}}{3.785 \text{ liters}} = \frac{20 \text{ gallons}}{L \text{ liters}}$$

Cross multiply.

$$L = 3.785 \times 20 = 75.7 \text{ liters per week}$$

Since 75.7 liters of water are wasted each week, $2 \times 75.7 =$ 151.4 liters of water are wasted in two weeks.

3. There are 16 ounces in a pound. Therefore, there are 16 ounces $\div$ 1.61 ounces $\approx$ 9.9 golf balls per pound. So, there are about 10 golf balls in one pound of golf balls. For the tennis balls, 16 ounces $\div$ 2.06 ounces $\approx$ 7.8 tennis balls. There are about 8 tennis balls in one pound of tennis balls.
There are about $10 - 8 = 2$ more golf balls in a pound than tennis balls in a pound.

4. To convert degrees Fahrenheit to degrees Celsius, use the formula from page 218.

$$C = (F - 32) \div 1.8$$

$$C = (49.1 - 32) \div 1.8$$

$$C = 17.1 \div 1.8 = 9.5° \text{ Celsius}$$

The maximum average daily temperature in Tokyo in the month of January is 9.5° Celsius.

5. A non-leap year has 365 days. At 0.004 inches per day, your toenails would grow 365 days $\times$ 0.004 inches per day = 1.46 inches in one year.

6. Change 5 dozen cookies to 5×12 cookies per dozen = 60 cookies.

   Use a proportion. Let c stand for the number of cups of flour in 160 cookies.

   $$\frac{2\frac{1}{4} \text{ cups}}{60 \text{ cookies}} = \frac{c \text{ cups}}{160 \text{ cookies}}$$

   Cross multiply.

   $$2\frac{1}{4} \times 160 = 60 \times c$$

   $$\frac{9}{\cancel{4}_{1}} \times \frac{\cancel{160}^{40}}{1} = 60 \times c$$

   $$360 = 60 \times c$$

   Divide both sides by 60.

   $$c = 360 \div 60 = 6$$

   To make 160 cookies, 6 cups of flour are needed.

7. Find all of the different routes. The route that measures 2.5 miles is from the school to the fire station (1.6 miles) to the library (0.9 mile): $1.6 + 0.9 = 2.5$ miles.

8. Using the different routes found in Problem 7, the shortest route is from the school to the store (0.5 mile) to the library (0.8 mile): $0.5 + 0.8 = 1.3$ miles.

9. Look at the conversion chart to find that one meter equals 39.37 inches. Change 10 feet to inches: 10 feet $\times$ 12 inches per foot = 120 inches.

   Use a proportion. Let m stand for the rim's height in meters.

   $$\frac{1 \text{ meter}}{39.37 \text{ inches}} = \frac{m \text{ meters}}{120 \text{ inches}}$$

   Cross multiply.

   $$120 = 39.37 \times m$$

   Divide both sides of the equation by 39.37.

$$120 \div 39.37 \approx 3.05 \text{ meters}$$

The rim is about 3.05 meters from the court surface.

Set # 22, page 225

1. The hypotenuse c (sign to shadow) is unknown.
    The legs a and b are 6 feet and 8 feet long (in either order).
    Use the Pythagorean Theorem.

    $$a^2 + b^2 = c^2$$
    $$6^2 + 8^2 = c^2$$
    $$36 + 64 = c^2$$
    $$c^2 = 100$$

    Since $10 \times 10 = 100$, $c = 10$ feet.
    The distance from the top of the sign to the tip of the shadow
    is 10 feet.

2. Draw a picture. The ladder forms the
    hypotenuse side c. The legs a and b are
    12 feet and 5 feet.
    Use the Pythagorean Theorem.

    $$a^2 + b^2 = c^2$$
    $$5^2 + 12^2 = c^2$$
    $$25 + 144 = c^2$$
    $$c^2 = 169$$

    Since $13 \times 13 = 169$, $c = 13$ feet.
    The ladder is 13 feet long.

3. Draw the top of a rectangular box.

    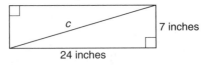

    The ribbon will form a diagonal of the rectangular top, or the
    hypotenuse of a right triangle.

Use the Pythagorean Theorem.

$$a^2 + b^2 = c^2$$

$$7^2 + 24^2 = c^2$$

$$49 + 576 = c^2$$

$$c^2 = 625$$

Since $25 \times 25 = 625$, $c = 25$ feet.
The ribbon should be 25 inches long.

4. Draw a picture.
   Use the Pythagorean Theorem.

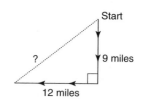

$$a^2 + b^2 = c^2$$

$$9^2 + 12^2 = c^2$$

$$81 + 144 = c^2$$

$$c^2 = 225$$

Since $15 \times 15 = 225$, $c = 15$ miles.
Sorange was 15 miles from her starting spot.

5. Draw a picture.
   Use the Pythagorean Theorem.

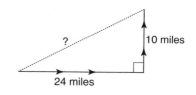

$$a^2 + b^2 = c^2$$

$$24^2 + 10^2 = c^2$$

$$576 + 100 = c^2$$

$$c^2 = 676$$

Since $26 \times 26 = 676$, $c = 26$.

The distance from Isaac's home to the concert hall is 26 miles.

CHAPTER EIGHT

Enriching Equations and a Bit of Algebra

Here are variables and equations
Our last strategy to see
How solving word problems
Can be done so painlessly!

SOLVING WORD PROBLEMS BY WRITING AND SOLVING LINEAR EQUATIONS

Did you know that you have already solved many algebra word problems in Chapters Two through Seven? A problem solved with an equation is an algebra problem. The unknown in the problem is expressed as a *variable* in the equation. Algebra problems can be solved by guessing and checking your guess, making a list, drawing a picture, or solving equations. This chapter will focus on solving linear equations in ways that you may not have already learned. It is wise to go back to using Polya's Four Steps of Problem Solving, relating the steps specifically to solving problems by solving equations.

MATH NOTE

A linear equation is an equation in which the unknown number or numbers appear only to the first power. Examples of linear equations are $3x + 5 = 14$ and $2A + B = 10$. The equation $5x^2 + 60 = 185$ is not a linear equation. Note that when there is no operation sign between a number and a symbol, the operation to be used is multiplication. Thus, $3x$ is a shortened form of $3 \times x$. If a number does not appear in front of a variable, a 1 is understood: $s = 1s = 1 \times s$

Polya's four steps: an effortless equation plan

Step 1: Understand the problem.

Read the problem carefully and ask yourself the following questions.

What is the problem about?
What is the unknown?
Is there enough information given to solve the problem?
Does the problem situation imply the operation of addition, subtraction, multiplication, or division?

Step 2: **Plan a strategy.**

Pick a letter or symbol to stand for the unknown number. (The letter or symbol is the variable. You can think of the variable as a mystery number!)

Write an equation using the letter or symbol, the given numbers, the operations you have identified, and math symbols.

Step 3: **Do the plan.**

Work backward to solve for the unknown. Working backward means reversing the order of the operations and using inverse operations. (Refer to Chapter Two for help with inverse operations and the working backward strategy.)

Step 4: **Check your work.**

Once you have found an answer, substitute your answer into the *original* equation. Perform all computations. If both sides of the resulting equation are equal, the answer is correct. If your answer does not check, try solving the problem again.

Follow the plan carefully
And you will soon find,
That solving an equation
Can be done in no time!

EXAMPLE:

Carlos bought 12 baseball cards, and now has 349 cards in his collection. How many cards did he have before his new purchase?

Step 1: Understand the problem.

Read the problem carefully and ask yourself the following questions.

What is the problem about? Carlos' baseball card collection

What is the unknown? how many cards Carlos had before he bought 12 more

Is there enough information given to solve the problem? Yes; we know the number of cards he bought and how many he now has.

Does the problem situation imply the operation of addition, subtraction, multiplication, or division? Addition is implied, since "bought" suggests adding more cards.

Step 2: Plan a strategy.

Pick a letter or symbol to stand for the number of cards Carlos had before he bought 12 more. Let c stand for the original number of cards.

Write an equation using the variable, the given numbers, the operations you have identified, and math symbols: $c + 12 = 349$.

Step 3: Do the plan.

The equation to be solved is $c + 12 = 349$.

There is only one operation, addition, involved in the equation.

The inverse of addition is subtraction.

Subtract 12 from both sides of the equation.

$$c + 12 - 12 = 349 - 12$$

$$c + 0 = 337$$

$$c = 337$$

Carlos had 337 baseball cards in his collection.

Step 4: Check your work.

Substitute the value you have found for the unknown into the original equation. This should result in an equality.

$$c + 12 \overset{?}{=} 349$$

$$337 + 12 \overset{?}{=} 349$$

$$349 = 349 \quad \checkmark$$

THE BALANCE SCALE MODEL

Finding the solution to a linear equation can be illustrated by the *balance scale model*. This model is illustrated below. For the scale to remain balanced, what is done on one side of the scale must be done on the other side. Similarly, when solving an equation, what is done on one side of the equation must be done on the other side of the equation in order to keep an equality. For

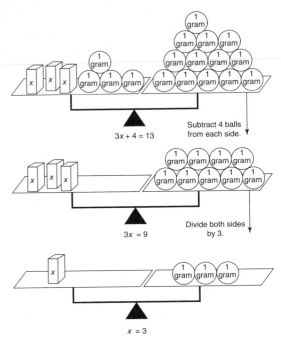

$3x + 4 = 13$

Subtract 4 balls from each side.

$3x = 9$

Divide both sides by 3.

$x = 3$

linear equations, you can do the following operations.

a. Add the same number to both sides of the equation.
b. Subtract the same number from both sides of the equation.
c. Multiply both sides by the same nonzero number.
d. Divide both sides by the same nonzero number.

EXAMPLE:

Twice Shyreen's age plus four years is Meredith's age. Meredith is 16 years old. Find Shyreen's age.

Step 1: Understand the problem.
Read the problem carefully and ask yourself the following questions.
What is the problem about? Shyreen's and Meredith's ages
What is the unknown? Shyreen's age
Is there enough information given to solve the problem?
Yes; we know the relationship between Shyreen's age and Meredith's age, and we know Meredith's age.

Does the problem situation imply the operation of addition, subtraction, multiplication, or division? "Twice" implies multiplication; "plus" implies addition.

Step 2: Plan a strategy. Pick a letter or symbol to stand for Shyreen's age. Let *s* stand for Shyreen's age. Write an equation using the variable, the given numbers, the operations you have identified, and math symbols: $2s + 4 = 16$. (Note that $2s$ means the same thing as $2 \times s$.)

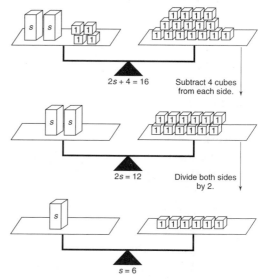

$2s + 4 = 16$ Subtract 4 cubes from each side.

$2s = 12$ Divide both sides by 2.

$s = 6$

Step 3: Do the plan.
The equation to be solved is $2s + 4 = 16$.
Work backward. The last computation done is addition.
The inverse of addition is subtraction.
Subtract 4 from both sides of the equation.

$$2s + 4 - 4 = 16 - 4$$
$$2s + 0 = 12$$
$$2s = 12$$

The inverse of multiplication is division.
Divide both sides of the equation by 2.

$$2s \div 2 = 12 \div 2$$
$$s = 6$$

Shyreen is six years old.

Step 4: Check your work.

Substitute the value you have found for the unknown into the original equation. This should result in an equality.

$$2s + 4 \stackrel{?}{=} 16$$
$$2 \times 6 + 4 \stackrel{?}{=} 16$$
$$12 + 4 \stackrel{?}{=} 16$$
$$16 = 16 \quad \checkmark$$

EXAMPLE:

Twice Dominic's age plus one equals three times his age less four. How old is Dominic now?

Step 1: Understand the problem.

Read the problem carefully and ask yourself the following questions.

What is the problem about? Dominic's age

What is the unknown? Dominic's current age

Is there enough information given to solve the problem? Yes; we know two equal conditions about Dominic's age.

Does the problem situation imply the operation of addition, subtraction, multiplication, or division? "Twice" and "three times" imply multiplication; "plus" implies addition; "less" implies subtraction.

Step 2: Plan a strategy.

Pick a letter or symbol to stand for Dominic's age. Let d stand for Dominic's age.

Write an equation using the variable, the given numbers, the operations you have identified, and math symbols: $2d + 1 = 3d - 4$.

Step 3: Do the plan.

The equation to be solved is $2d + 1 = 3d - 4$.

Notice that there are variables on both sides of the equation. Plan to get the variables on the same side of the equation and the numbers on the other side.

Subtract $2d$ from both sides of the equation.

$$2d + 1 - 2d = 3d - 4 - 2d$$
$$1 = d - 4$$

Add 4 to both sides of the equation.

$$1 + 4 = d - 4 + 4$$
$$5 = d$$
$$d = 5$$

Dominic is five years old.

Step 4: Check your work.
Substitute the value you have found for the unknown into the original equation. This should result in an equality.

$$2d + 1 \overset{?}{=} 3d - 4$$
$$2 \times 5 + 1 \overset{?}{=} 3 \times 5 - 4$$
$$10 + 1 \overset{?}{=} 15 - 4$$
$$11 = 11 \quad \checkmark$$

MATH NOTE

When there are variables and numbers on both sides of an equation, plan to end with the variables on one side and the numbers on the other side. In the previous example, the numbers ended up on the left side and the variable ended up on the right side. In an equality, the left and right sides of the equation can be interchanged.

EXAMPLE:

The sum of an even number and the next consecutive even number is 54. Find the smaller of the consecutive even numbers.

Step 1: Understand the problem.
Read the problem carefully and ask yourself the following questions.

What is the problem about? the sum of two consecutive even numbers

What is the unknown? the smaller of the two even numbers

Is there enough information given to solve the problem? Yes; we know that consecutive even numbers differ by two, and we know the sum of the two consecutive even numbers.

Does the problem situation imply the operation of addition, subtraction, multiplication, or division? "Sum" implies addition.

Step 2: Plan a strategy.

Pick a letter or symbol to stand for the smaller of the two even numbers. Let e stand for the smaller even number. Then $e + 2$ stands for the next even number. Write an equation using the variable, the given numbers, the operations you have identified, and math symbols: $e + (e + 2) = 54$.

Step 3: Do the plan.

The equation to be solved is $e + (e + 2) = 54$.

$$e + (e + 2) = 54$$
$$(e + e) + 2 = 54$$
$$2e + 2 = 54$$

Subtract 2 from both sides of the equation.

$$2e + 2 - 2 = 54 - 2$$
$$2e = 52$$

Divide both sides of the equation by 2.

$$2e \div 2 = 52 \div 2$$
$$e = 26, \text{ and so } e + 2 = 28$$

The smaller of the two consecutive even numbers is 26.

Step 4: Check your work.

Substitute the value you have found for the unknown into the original equation. This should result in an equality.

$$e + (e + 2) \overset{?}{=} 54$$

$$26 + (26 + 2) \overset{?}{=} 54$$

$$26 + 28 \overset{?}{=} 54$$

$$54 = 54 \quad \checkmark$$

MATH NOTE

Algebra problems sometimes involve finding *consecutive* numbers, numbers that directly follow each other. Since whole numbers go up by one as you count them (1, 2, 3, 4, and so on), the variables for consecutive whole numbers are x and $x + 1$. Since even and odd numbers go up by two as you count them (0, 2, 4, 6, . . . or 1, 3, 5, 7, . . .), the variables for consecutive even or odd numbers are x and $x + 2$.

BRAIN TICKLERS
Set # 23

Read each problem. Pick a letter or symbol to stand for the unknown. Write an equation using the letter or symbol, the given numbers, the operations you have identified, and math symbols. Then solve the equation. Of course, check your work.

At the end of each problem you will find a letter. By placing this letter on the line above the corresponding answer in the answer code at the end of the problem set, you will be able to answer this question:

Who invented the potato chip?

1. Shamika gave $129 of her monthly take-home pay to charity. If she was left with $2,479, what is her take-home pay? (O)

2. Drake Dodger had 138 hits this season, 19 more hits than last season. How many hits did he have last season? (R)

3. This year Marlenny Mulch grew four more than twice the number of tomato plants she grew last year. This year she planted a total of 88 tomato plants. How many tomato plants did she grow last year? (E)

4. The seventh grade students who went on the field trip were divided into four equal groups to ride on four buses. Twelve of the seventh grade students did not go on the field trip. There was a total of 38 students on each bus. How many students are in the seventh grade? (U)

5. The sum of a whole number and the next consecutive whole number is 37. Find the smaller of the two numbers. (G)

6. The sum of an odd number and the next consecutive odd number is 24. Find the larger of the two numbers. (G)

7. The number of students at Weeks Middle School is five less than four times the number of students at Bicentennial Middle School. There are 491 students at Weeks School. How many students attend Bicentennial School? (M)

8. In seven years Jovan will be twice her brother's present age. Her brother is 14 years old. How old is Jovan now? (R)

9. Hugh divides his monthly salary into four equal parts for his budget. He saves $50 less than one equal part each month. Each month he saves $605. What is his monthly salary? (C)

10. I am thinking of a mystery number. Three more than six times the number is equal to three times the number, plus nine. What's my mystery number? (E)

Answer Code _____ _____ _____ _____ _____ _____
                 18   2   2,608   21   13   42

_____ _____ _____ _____ invented the potato chip.
2,620 119   164   124

(Answers are on page 254.)

WORD PROBLEMS SOLVED BY SOLVING TWO LINEAR EQUATIONS

Sometimes there are two unknowns in a problem. In this case, two equations are needed for its solution. Each unknown is given a variable name, then two equations are written using the variables, the given numbers, the operations implied, and math symbols. To solve the problem, you will need to learn how to solve two equations with two unknowns.

> But please rest assured,
> Whether it's one equation or two,
> Algebra problems will be painless for you.

EXAMPLE:

Xeno bought a CD and a tape. Without the tax, the total cost was $20. The CD cost $10 more than the tape. Find the cost of each item.

Step 1: Understand the problem.

Read the problem carefully and ask yourself the following questions.

What is the problem about? the prices of a CD and a tape

What are the unknowns? the price of the CD and the price of the tape

Is there enough information given to solve the problem? Yes; we know the total cost of the two items, and how much more the CD costs than the tape.

Does the problem situation imply the operation of addition, subtraction, multiplication, or division? "Total cost" implies addition; "more than" also implies addition.

Step 2: Plan a strategy.

Pick a letter or symbol to stand for each of the two items. Let c stand for the cost of the CD, and let t stand for the cost of the tape.

Write two equations using the variables, the given numbers, the operations you have identified, and math symbols.

Equation One: The cost of the CD plus the cost of the tape equals $20, so $c + t = 20$.

Equation Two: The CD cost $10 more than the tape, so $c = 10 + t$.

Step 3: Do the plan.

There are two equations to be solved:

$$c + t = 20 \quad \text{and}$$

$$c = 10 + t$$

Since we know from Equation Two that c equals $10 + t$, substitute $10 + t$ for c in Equation One.

$$c + t = 20$$

$$(10 + t) + t = 20$$

$$10 + (t + t) = 20$$

$$10 + 2t = 20$$

Subtract 10 from both sides of the equation.

$$2t + 10 - 10 = 20 - 10$$

$$2t + 0 = 10$$

$$2t = 10$$

Divide each side by 2.

$$2t \div 2 = 10 \div 2$$

$$t = 5$$

The tape cost $5. To find the cost of the CD, substitute 5 for t in Equation One.

$$c + 5 = 20$$

Subtract 5 from both sides of the equation.

$$c + 5 - 5 = 20 - 5$$
$$c + 0 = 15$$
$$c = 15$$

The tape cost $5 and the CD cost $15.

Step 4: Check your work.
Do the two prices add up to $20? $15 + $5 = $20 ✓
Does the CD cost $10 more than the tape?
$15 − $5 = $10 ✓

EXAMPLE:

At a movie theater, popcorn costs $5 less than a ticket. The total cost of both items is $11. Find the cost of the ticket.

Step 1: Understand the problem.
Read the problem carefully and ask yourself the following questions.
What is the problem about? the cost of a movie ticket and popcorn
What are the unknowns? the cost of the movie ticket and the cost of the popcorn
Is there enough information given to solve the problem? Yes; we know the total cost of the two items, and how much less the popcorn costs than the movie ticket.
Does the problem situation imply the operation of addition, subtraction, multiplication, or division? "Total cost" implies addition; "less than" implies subtraction.

Step 2: Plan a strategy.
Pick a letter or symbol to stand for each of the two items. Let p stand for the cost of the popcorn, and let t stand for the cost of the movie ticket.
Write two equations using the variables, the given numbers, the operations you have identified, and math symbols.

Equation One: The cost of both items is \$11, so $t + p = 11$.

Equation Two: The popcorn costs \$5 less than a ticket, so $p = t - 5$.

Step 3: Do the plan.

There are two equations to be solved:

$$t + p = 11 \quad \text{and}$$

$$p = t - 5$$

Since we know from Equation Two that p equals $t - 5$, substitute $t - 5$ for p in Equation One.

$$t + p = 11$$
$$t + (t - 5) = 11$$
$$(t + t) - 5 = 11$$
$$2t - 5 = 11$$

Add 5 to both sides of the equation.

$$2t - 5 + 5 = 11 + 5$$
$$2t + 0 = 16$$
$$2t = 16$$

Divide each side by 2.

$$2t \div 2 = 16 \div 2$$
$$t = 8$$

The ticket costs \$8. To find the cost of the popcorn, substitute 8 for t in Equation One.

$$8 + p = 11$$

Subtract 8 from both sides of the equation.

$$8 - 8 + p = 11 - 8$$
$$p = 3$$

The ticket costs \$8 and the popcorn costs \$3.

Step 4: Check your work.
Do the two prices add up to $11? $8 + $3 = $11 ✓
Does the popcorn cost $5 less than the ticket?
$8 − $5 = $3 ✓

MATH NOTE

With two equations, remember: Substitute the value for one variable in one equation into its spot in the other equation. Solve and check your work. This method is known as the *method of substitution*.

BRAIN TICKLERS
Set # 24

Read each problem. Plan a strategy by looking for the two unknowns and writing two equations. Do the plan using substitution. Check your work.

1. A drink and a slice of pizza costs a total of $3.50. The pizza costs $1.50 more than the drink. Find the cost of the drink.

2. There is a total of 300 boys and girls at Emanuel School. There are 60 more girls than boys at the school. How many girls attend Emanuel School?

3. Cynthia is three years older than Beverly. The sum of their ages is 81. How old is Cynthia?

4. Charles is 10 years younger than Stanley. The sum of their ages is 90. Find Charles' age.

5. A pair of shoes costs $10 more than a pair of sneakers. Together, their cost is $110. Find the cost of the sneakers.

6. The sum of two numbers is 20. One number is six more than the other number. Find the larger number.

7. The sum of two numbers is 35. One number is five less than the other number. Find the smaller number.

8. The number of calories in a muffin is 50 more than twice the number of calories in a cookie. Together they have 350 calories. How many calories are in a cookie?

9. There is a total of 31 kinds of burgers and sandwiches on a menu. The sandwiches total five more than the burgers. How many kinds of burgers are on the menu?

10. The sum of two numbers is 37. The larger number is one more than twice the smaller number. Find the smaller number.

(Answers are on page 260.)

This chapter has presented an introduction to solving linear problems with one or two unknowns. To learn more about solving equations, see Long, Lynnette. *Painless Algebra*. Barron's Educational Series, Inc., 1998.

BRAIN TICKLERS—THE ANSWERS

Set # 23, page 247

The variables that you use to solve the problems may be different from the ones stated below. That's all right! Just make sure that your answers match the answers stated.

1. The unknown number is Shamika's take-home pay. Let e stand for her take-home pay. Her take-home pay less her $129 charity contribution is the amount she has left.

$$t - \$129 = \$2,479$$

Add 129 to both sides of the equation.

$$t - 129 + 129 = 2,479 + 129$$
$$t + 0 = 2,608$$
$$t = 2,608$$

Shamika's take-home pay is $2,608.
Put an O on the line above 2,608.

2. The unknown is the number of hits Drake had last season.
   Let h stand for the number of hits last season. The number of
   hits this season is 19 more than the number last season. This
   season he made a total of 138 hits.

$$h + 19 = 138$$

Subtract 19 from both sides.

$$h + 19 - 19 = 138 - 19$$
$$h + 0 = 119$$
$$h = 119$$

Drake made 119 hits last season.
Put an R on the line above 119.

3. The unknown is the number of tomato plants Marlenny grew
   last year. Let p stand for the number of tomato plants she
   grew last year. This year she grew four more than twice the
   number of tomato plants she grew last year.

$$2p + 4 = 88$$

Subtract 4 from both sides of the equation.

$$2p + 4 - 4 = 88 - 4$$
$$2p + 0 = 84$$
$$2p = 84$$

Divide each side of the equation by two.

$$2p \div 2 = 84 \div 2$$

$$p = 42$$

Marlenny grew 42 tomato plants last year.
Put an E on the line above 42.

4. The unknown number is the total number of seventh grade
    students. Let s stand for the total number of students. The
    number of students who went on the trip is 12 less than the
    total number of students, so $s - 12$ students went on the trip.
    These students were divided into four equal groups. There
    were 38 students in each group.

$$(s - 12) \div 4 = 38$$

Write the left-hand side as a fraction.

$$\frac{s - 12}{4} = 38$$

Multiply both sides of the equation by 4.

$$\frac{s - 12}{4} \times 4 = 38 \times 4$$

$$\frac{s - 12}{\cancel{4}^{1}} \times \frac{\cancel{4}^{1}}{1} = 152$$

$$s - 12 = 152$$

Add 12 to both sides of the equation.

$$s - 12 + 12 = 152 + 12$$

$$s + 0 = 164$$

$$s = 164$$

There are 164 students in the seventh grade.
Put a U on the line above 164.

5. There are two unknown numbers. Let x be the first whole
    number. Then $x + 1$ is the next consecutive whole number.
    (Remember that *consecutive* means one after the other, and

that whole numbers go up by one.) The known number is the sum, 37. The sum of the two consecutive numbers is 137.

$$x + (x + 1) = 37$$
$$(x + x) + 1 = 37$$
$$2x + 1 = 37$$

Subtract 1 from both sides of the equation.

$$2x + 1 - 1 = 37 - 1$$
$$2x + 0 = 36$$
$$2x = 36$$

Divide each side of the equation by 2.

$$2x \div 2 = 36 \div 2$$
$$x = 18$$

The first number, or the smallest number, is 18.
Put a G on the line above 18.

6. There are two unknown numbers, the first odd number (x) and the next consecutive odd number $(x + 2)$. (Remember that odd numbers go up by two.) The sum of the two consecutive odd numbers is 24.

$$x + (x + 2) = 24$$
$$(x + x) + 2 = 24$$
$$2x + 2 = 24$$

Subtract 2 from both sides of the equation.

$$2x + 2 - 2 = 24 - 2$$
$$2x + 0 = 22$$
$$2x = 22$$
$$x = 11, \text{ and so } x + 2 = 13$$

The larger of the two consecutive odd numbers is 13.
Put a G on the line above 13.

7. The unknown number is the number of students at Bicentennial Middle School (b). The number of students at Weeks Middle School is five less than four times the number of students at Bicentennial. There are 491 students at Weeks.

$$4b - 5 = 491$$

Add 5 to both sides of the equation.

$$4b - 5 + 5 = 491 + 5$$

$$4b + 0 = 496$$

$$4b = 496$$

Divide both sides of the equation by 4.

$$4b \div 4 = 496 \div 4$$

$$b = 124$$

There are 124 students at Bicentennial Middle School.
Put an M on the line above 124.

8. The unknown number is Jovan's current age (j). In 7 years Jovan will be twice her brother's present age of 14.

$$j + 7 = 2 \times 14$$

$$j + 7 = 28$$

Subtract 7 from both sides of the equation.

$$j + 7 - 7 = 28 - 7$$

$$j + 0 = 21$$

$$j = 21$$

Jovan is 21 years old.
Put an R on the line above 21.

9. The unknown number is Hugh's monthly salary (s). He divides his salary into four equal parts. He saves $50 less than one equal part each month. He saves $605 each month.

$$\frac{s}{4} - 50 = 605$$

Add 50 to both sides of the equation.

$$\frac{s}{4} - 50 + 50 = 605 + 50$$

$$\frac{s}{4} + 0 = 655$$

$$\frac{s}{4} = 655$$

Multiply each side of the equation by 4.

$$^{1}\cancel{4} \times \frac{s}{\cancel{4}_1} = 4 \times 655$$

$$\frac{4}{1} \times \frac{3}{4} = 2{,}620$$

$$s = 2{,}620$$

Hugh's monthly salary is $2,620.
Put a C on the line above 2,620.

10. The unknown number is a mystery number (x). Three more than six times the number ($6 \times x + 3$) is equal to three times the number, plus nine ($3 \times x + 9$).

$$6x + 3 = 3x + 9$$

Subtract 3 from both sides of the equation.

$$6x + 3 - 3 = 3x + 9 - 3$$

$$6x + 0 = 3x + 6$$

$$6x = 3x + 6$$

Subtract $3x$ from both sides of the equation.

$$6x - 3x = 3x - 3x + 6$$

$$3x = 0 + 6$$

$$3x = 6$$

Divide each side of the equation by 3.

$$3x \div 3 = 6 \div 3$$

$$x = 2$$

The mystery number is 2.

Put an E on the line above 2.

Answer Code	G	E	O	R	G	E
	18	2	2,608	21	13	42

C	R	U	M	
2,620	119	164	124	invented the potato chip.

Set # 24, page 253

1. The total cost of the drink (d) and the slice of pizza (p) is $3.50. The slice of pizza costs $1.50 more than the drink.

$$\text{Equation One: } d + p = 3.50$$

$$\text{Equation Two: } p = 1.50 + d$$

Substitute $1.50 + d$ for p in Equation One.

$$d + (1.50 + d) = 3.50$$
$$(d + d) + 1.50 = 3.50$$
$$2d + 1.50 = 3.50$$

Subtract 1.50 from both sides of the equation.

$$2d + 1.50 - 1.50 = 3.50 - 1.50$$
$$2d + 0 = 2.00$$
$$2d = 2.00$$

Divide both sides of the equation by 2.

$$2d \div 2 = 2.00 \div 2$$
$$d = 1.00$$

The drink costs $1.00.

2. The total number of boys (b) and girls (g) at Emanuel School is 300. There are 60 more girls than boys.

$$\text{Equation One: } b + g = 300$$

$$\text{Equation Two: } 60 + b = g$$

Substitute $60 + b$ for g in Equation One.

$$b + (60 + b) = 300$$

$$(b + b) + 60 = 300$$

$$2b + 60 = 300$$

Subtract 60 from both sides of the equation.

$$2b + 60 - 60 = 300 - 60$$

$$2b + 0 = 240$$

$$2b = 240$$

Divide both sides of the equation by 2.

$$2b \div 2 = 240 \div 2$$

$$b = 120$$

There are 120 boys at Emanuel School.
Therefore, there are $300 - 120 = 180$ girls at the school.

3. Cynthia is three years older than Beverly. The sum of Cynthia's age (c) and Beverly's age (b) is 81.

$$\text{Equation One: } c = 3 + b$$

$$\text{Equation Two: } c + b = 81$$

Substitute $3 + b$ for c in Equation Two.

$$(3 + b) + b = 81$$

$$3 + (b + b) = 81$$

$$3 + 2b = 81$$

Subtract 3 from both sides of the equation.

$$3 + 2b - 3 = 81 - 3$$

$$2b + 0 = 78$$

$$2b = 78$$

Divide both sides of the equation by 2.

$$2b \div 2 = 78 \div 2$$

$$b = 39$$

Beverly is 39 years old.
Since Beverly is 39 years old, Cynthia is $39 + 3 = 42$ years old.

4. Charles is ten years younger than Stanley. The sum of Charles' age (c) and Stanley's age (s) is 90.

Equation One: $c = s - 10$

Equation Two: $c + s = 90$

Substitute $s - 10$ for c in Equation Two.

$$(s - 10) + s = 90$$

$$(s + s) - 10 = 90$$

$$2s - 10 = 90$$

Add 10 to both sides of the equation.

$$2s - 10 + 10 = 90 + 10$$

$$2s + 0 = 100$$

$$2s = 100$$

Divide both sides of the equation by 2.

$$2s \div 2 = 50 \div 2$$

$$s = 50$$

Stanley is 50 years old.
Since Stanley is 50 years old, Charles is $50 - 10 = 40$ years old.

5. A pair of shoes costs $10 more than a pair of sneakers. The total cost of the shoes (s) and the sneakers (k) is $110.

$$\text{Equation One: } s = 10 + k$$

$$\text{Equation Two: } k + s = 110$$

Substitute $10 + k$ for s in Equation Two.

$$k + (10 + k) = 110$$

$$(k + k) + 10 = 110$$

$$2k + 10 = 110$$

Subtract 10 from both sides of the equation.

$$2k + 10 - 10 = 110 - 10$$

$$2k + 0 = 100$$

$$2k = 100$$

Divide both sides of the equation by 2.

$$2k \div 2 = 100 \div 2$$

$$k = 50$$

The sneakers cost $50.

6. Let x stand for the larger number and y for the smaller number. The sum of the numbers is 20. One number is six more than the other.

$$\text{Equation One: } x + y = 20$$

$$\text{Equation Two: } x = 6 + y$$

Substitute $6 + y$ for x in Equation One.

$$(6 + y) + y = 20$$

$$6 + (y + y) = 20$$

$$6 + 2y = 20$$

Subtract 6 from both sides of the equation.

$$6 + 2y - 6 = 20 - 6$$

$$2y + 0 = 14$$

$$2y = 14$$

Divide both sides of the equation by 2.

$$2y \div 2 = 14 \div 2$$

$$y = 7$$

The smaller number is 7.
The larger number is $20 - 7 = 13$.

7. Let x stand for the larger number and y for the smaller number. The sum of the numbers is 35. One number is five less than the other.

$$\text{Equation One: } x + y = 35$$

$$\text{Equation Two: } y = x - 5$$

Substitute $x - 5$ for y in Equation One.

$$x + (x - 5) = 35$$

$$(x + x) - 5 = 35$$

$$2x - 5 = 35$$

Add 5 to both sides of the equation.

$$2x - 5 + 5 = 35 + 5$$

$$2x + 0 = 40$$

$$2x = 40$$

Divide both sides of the equation by two.

$$2x \div 2 = 40 \div 2$$

$$x = 20$$

The larger number is 20.
The smaller number $20 - 5 = 15$.

8. The number of calories in a muffin (m) is 50 more than twice the number of calories in a cookie (c). The total number of calories in both the muffin and the cookie is 350.

$$\text{Equation One: } m = 50 + 2c$$

$$\text{Equation Two: } m + c = 350$$

Substitute $50 + 2c$ for m in Equation Two.

$$(50 + 2c) + c = 350$$

$$50 + (2c + c) = 350$$

$$50 + 3c = 350$$

Subtract 50 from both sides of the equation.

$$50 + 3c - 50 = 350 - 50$$

$$3c + 0 = 300$$

$$3c = 300$$

Divide both sides of the equation by 3.

$$3c \div 3 = 300 \div 3$$

$$c = 100$$

There are 100 calories in a cookie.

9. Let s stand for the number of sandwiches and b for the number of burgers. There are 31 kinds of burgers and sandwiches. The number of sandwiches is 5 more than the number of burgers.

$$\text{Equation One: } s + b = 31$$

$$\text{Equation Two: } s = b + 5$$

Substitute $b + 5$ for s in Equation One.

$$(b + 5) + b = 31$$

$$(b + b) + 5 = 31$$

$$2b + 5 = 31$$

Subtract 5 from both sides of the equation.

$$2b + 5 - 5 = 31 - 5$$

$$2b + 0 = 26$$

$$2b = 26$$

Divide both sides of the equation by two.

$$2b \div 2 = 26 \div 2$$

$$2b = 13$$

There are 13 burgers.

10. Let x stand for the smaller number. Let y stand for the larger number. The sum of the two numbers is 37. The larger number is one more than twice the smaller number.

Equation One: $x + y = 37$

Equation Two: $y = 1 + 2x$

Substitute $1 + 2x$ for y in Equation One.

$$x + (1 + 2x) = 37$$

$$(x + 2x) + 1 = 37$$

$$3x + 1 = 37$$

Subtract 1 from both sides of the equation.

$$3x + 1 - 1 = 37 - 1$$

$$3x + 0 = 36$$

$$3x = 36$$

Divide both sides of the equation by 3.

$$3x \div 3 = 36 \div 3$$

$$x = 12$$

Twelve is the smaller number.

CHAPTER NINE

A Potpourri of Practice Problems

From whole numbers to algebra,
With many topics in between,
You've learned many strategies—
So try some for practice and join the *Painless* team!

1. Kylie has 18 more CDs than Kira. Kira has 17 fewer CDs than Ken. Ken has double the number of CDs of Kora. Kora has 16 CDs. How many CDs does Kylie have?

2. The two top prizes at the talent show are tickets to a concert. The four finalists are Jan, Julie, Jim, and Jerry. If two of the finalists win tickets, how many different combinations of two winners are possible?

3. Sammy Saver had a balance of $100 in his checking account. He then withdrew $29, deposited $254, wrote a check for $17.95 and finally, deposited $85. What was Sammy's new balance?

4. A rectangular room, measuring 9 feet by 12 feet, will be covered with 6 inch by 6 inch square tiles. How many tiles will be needed to cover the entire floor of the room?

5. A circular swimming pool measures 50 feet in diameter. How many times would you have to swim around the pool to swim a distance of one mile?

6. Anna has test scores of 85, 75, 89, and 71. What is the least possible score that she can achieve on her next test to achieve an average of 82?

7. A recipe that serves four people requires $1\frac{1}{4}$ cups of flour. How many cups of flour will be needed in order for the recipe to serve 18 people?

8. Howie has three less than twice the number of books that Hortense has. If Howie has 21 books, how many books does Hortense have?

9. ThuVan, Gordon, and Lilizita each had 120 tickets to sell for the school raffle. ThuVan sold 25% of his tickets, Gordon sold $\frac{3}{8}$ of his tickets, and Lilizita sold $\frac{4}{15}$ of her tickets. How many tickets did the person who sold the most tickets sell?

10. In order to win a game, Kelsey needs to toss a sum of six using two dice. What is the probability that he will toss a sum of six on his next throw?

11. A pasta recipe that will serve eight people requires $\frac{3}{4}$ cup of olive oil. How many pints of oil will be needed if the recipe is to serve 40 people?

12. A pair of pants and a jacket cost a total of $200. The jacket costs $40 more than the pants. Find the cost of the jacket.

13. Film Time movie theater offers a coupon book of five tickets for $18.75. Reel House offers four tickets for $16.60. Oscar Palace offers six tickets for $23.10. Which theater offers the best discount rate, and what is the best rate?

14. To train for the 26-mile road race, Benita has planned a weekly practice schedule. She will run one mile each day for the first week, four miles each day for the second week, seven miles each day for the third week, and so on. When she reaches a distance greater than 15 she will only run twice a week. During which week will she first start to run at least as many miles as in the road race?

15. At Price Counter grocery store, crackers are three boxes for $5.25, corn chips are two bags for $1.80, and fruit juice costs $3.75 for three cartons. What is the cost of buying two boxes of crackers, one bag of corn chips, and two cartons of juice?

16. On Monday, one share of Microchip stock cost $75 $\frac{1}{2}$. During the week, the stock went up $\frac{1}{4}$ point, went down $1\frac{1}{8}$ points, and then went up $\frac{3}{8}$ point. At the end of the week, what was the cost of one share of Microchip stock?

17. Jordan's Basement store offers the following discount plan:

Week One: regular price

Week Two: 5% off of the regular price

Week Three: 15% off of week two's price

Week Four: 20% off of week three's price

What will be the cost of a sweater, regularly priced at $50.00, that is purchased during week four?

18. In Monday night's game, the four top-scoring players on the Hoopster basketball team had the following point totals:

Player	Points Scored
Baxter	21
Ha	12
Johnson	25
Lopez	8

What was the median number of points scored by these players?

19. A necklace that sells for $48 costs $51 with the tax included. Find the percent rate of the tax.

20. Cindy-Jo is buying wallpaper that costs $7.50 per square yard for a rectangular wall that measures 10 feet by 20 feet. Melissa is buying wallpaper that costs $6.00 per square yard for a rectangular wall measuring 12 feet by 18 feet. Who will spend the greatest amount and, to the nearest cent, how much more will they spend?

(Answers follow.)

PRACTICE PROBLEMS—THE ANSWERS

1. 33 CDs

2. 6 combinations

3. $392.05

4. 432 tiles

5. about 34 times

6. 90

7. $5\frac{5}{8}$ cups

8. 12 books

9. 45 tickets

10. $\frac{5}{36}$

11. $1\frac{7}{8}$ pints

12. $120.00

13. Film Time; $3.75 per ticket

14. the tenth week

15. $6.90

16. $75.00

17. $32.30

18. 16.5 points

19. $6\frac{1}{4}\%$

20. Cindy-Jo; $22.67

CHAPTER TEN

Interesting Internet Ideas

From whole numbers to algebra,
Word problems you now can do;
Here come some Internet activities—
Practice and projects just for you.

The Internet is a river of resources for practicing and solving all kinds of real-life word problems. There are World Wide Web sites for practicing the computation needed to solve word problems and/or to challenge you with curious and interesting problems. Using search techniques, you can find web sites that provide information and data on many areas and problems of your own interest and see intriguing charts, graphs, lists, pictures, and of course, numbers, in the process.

Internet Idea #1

For practice with computation with whole numbers, fractions, decimals, and/or basic problems in measurement and geometry, try this:

 http://quia.com

1. Click on **Mathematics**.

2. Choose a topic from either **Geometric Terms**, **Quiz on Equations**, **Fraction Decimal Conversion**, **Customary Measurement**, or **Square Roots**.

3. Play and practice away!

Internet Idea #2

For projects that you can do on your own or with your family, try this:

 http://www.figurethis.org

1. a. Click on **Challenge Index**.
   b. Scroll down and click on **Stamps**.
   c. Read the activity.
   d. Use a guess and check strategy or make a list to find all of the possible combinations of stamps that will fit the problem.

275

2. a. Click back to the **Challenge Index**.
    b. Scroll down and click on **Batting Averages**.
    c. Use the skills that you learned in solving word problems with ratios and proportions and statistics to find the best player.
3. a. Click back to the **Challenge Index**.
    b. Scroll down and click on **Play Ball!**
    c. Make a list or create an equation to decide whose baseball is worth the most money.
4. a. Click back to the **Challenge Index**.
    b. Scroll down and click on **Beating Heart**.
    c. Work with a friend or family member to measure your pulse. Use your Painless Word Problem skills to solve the heartbeat problem. Were you surprised at the answer?

Internet Idea #3

For activities that show some applications of math in everyday life, try this:

 http://www.learner.org/exhibits/dailymath/

1. a. Read the page, which describes how math is used in everyday life.
    b. Click on **Cooking by Numbers**.
    c. Read how ratios and proportions are used in cooking. Work along with the sample situation.
    d. Choose one of your favorite recipes. Increase or decrease the ingredients so that the recipe will serve all of the students in your school or just the students in your homeroom.
2. a. Click back to the original page and click on **Savings and Credit**.
    b. Read about simple and compound interest.
    c. Visit a local bank and make an appointment to talk to a manager or customer service person about interest and investments.
3. a. Click back to the original page and click on **Home Decorating**.

b. Read how geometry and measurement are used in finding the area of rooms in a floor plan. Work along with the sample situations.

c. Create a floor plan of your home, excluding closets or cabinet areas. Use a scale of one foot (actual) equals one inch (floor plan). Find the area of the rooms.

Internet Idea #4

For an interesting look at how measurement conversions are used in science, try this:

 http://www.exploratorium.edu/ronh/weight/index.html

1. a. Enter your weight in the box.
    b. Click on **Calculate**.
    c. Discover how much you would weigh on different planets.
    d. Use proportions and the data in the chart to calculate the approximate number of pounds or portion of a pound on different planets equal to one pound on planet Earth.

Internet Idea #5

For more practice in using algebraic equations to solve word problems, try this:

 http://www.about.com

1. a. Click on **Science**.
    b. Click on **Mathematics**.
    c. Look to the right near the top of the page for *Did You Know?* Click on **Unlock the secret to story problems!**
    d. Read the problems and the explanations of their solutions.
    e. Try some of the examples.

Internet Idea #6

For interactive and innovative activities, try this:

> **http://www.mathgoodies.com**

1. a. Under *Lessons*, click on **Topics in Pre-Algebra**.
   b. Click on **Writing Algebraic Expressions**.
   c. Read through the lesson.
   d. Try the Exercises.
   e. Click back to the original page, and under *Lessons*, click on **Understanding Percent**.
   f. Search through the lesson topics and click on one that interests you.
   g. Read through the lesson and try the Exercises.
   h. Click back to the original page, and under *Lessons*, choose from **Probability**, **Circumference & Area of Circles**, or **Perimeter & Area of Polygons**. Explore!

   > So many activities from which to choose, using
   > Painless strategies you cannot lose!

Internet Idea #7

For easy strategies and quick solutions to basic word problems try this:

> **http://www.webmath.com/**

1. a. Click on **Story Problems**.
   b. Choose a topic and click on either: **Addition Story Problems**, **Subtraction Story Problems**, **Multiplication Story Problems** or **Division Story Problems**.
   c. Read and solve all sorts of basic problems that rely on key words and computation.
   d. Return to the original page and click on **Real World Math**. Click on **Personal Finance**.
   e. Read about *simple interest* and *compound interest*, as well as other forms of saving and investing.
   f. Return to the original page and click on **Real World Math**.

g. Choose any of the following: under *Practical Math*, **Tips**; under *Unit Conversion*, **Length**, **Mass**, **Volume**, or **Temperature**.

Internet Idea #8

For activities that link mathematics with aeronautics engineering, try this:

http://www.planemath.com

1. a. Click on **Students**.
   b. Click on **Activities for Students**.
   c. Click on **Applying Flying**.
   d. Click on **Flight Path**.
   e. Click on **Lesson**.
   f. Read and follow the lesson to find the shortest flight path between the given cities.

Internet Idea #9

For a look at challenges and activities that are geared to seventh and eighth graders, check out *Mathcounts*, a national mathematics competition. Find it at:

http://www.mathcounts.org

1. a. Click on **Problem Solving**.
   b. Try some of the **Go Figure! Math Challenge** problems, which require the use of your Painless Word Problem skills and strategies. Or, you might want to explore **Problem Solving Strategies**.

Internet Idea #10

For games and activities that will help you to practice your basic mathematics skills try this:

http://www.funbrain.com/index.html

1. a. Click on your age.
    b. Choose and click on the mathematical skills that you would like to practice.
    c. Have fun practicing your math skills in a game format!

Internet Idea #11

For an interesting look at nutritional data and statistics at common fast-food restaurants try this:

http://www.cyberdiet.com

1. a. Look for *Diet and Nutrition* and click on **Fast Food Quest**.
    b. Under restaurants, click **McDonald's** and **Burger King**. (Un-click any other restaurants that are already checked.)
    c. Under *Fast Food Categories and Subcategories*, click **Hamburgers**.
    d. Under *Columns to Display*, un-click **Fat (%)**. Leave only the boxes for **Calories** and **Fat Grams** as checked boxes.
    e. Click on **Display Results**.
    f. Look at the data displayed. Find and compare the percent of fat in various hamburgers. Use the ratio of 1 fat gram = 9 calories. Use proportions to find the results.
    g. Click back and go to *Columns to Display*. Click **Fat (%)**.
    h. Click on **Display Results** to check your work.
    i. Were you surprised at the percentage of fat in hamburgers? Click back to compare the foods at other fast-food restaurants.

Internet Idea #12

Have you ever tried a frozen drink? Many frozen drinks have as many calories as solid foods. For an interesting look at nutrition and how statistics and proportions are used in a famous business, try this:

http://www.dunkindonuts.com

1. a. Click on **Nutritional Info**.
   b. Click on **Beverages**.
   c. Look and compare the calories and percentage of fat in Coffee Coolattas with cream, milk, 2% milk, and skim milk.
   d. How do the percentages of fat in Coffee Coolattas compare to those of the hamburgers from Internet Idea #11? Are you surprised?

Internet Idea #13

For a new problem each week that deals with geometry try this:

http://www.forum.swarthmore.edu/geopow/

Internet Idea #14

Do you have a mathematical question that has not been addressed in this book? Then *Ask Dr. Math* at:

http://www.forum.swarthmore.edu/dr.math/index.html

1. a. Click on **Middle School**.
   b. Near the bottom of the page, click on **Submit your own question to Dr. Math**, and enter your own question.
   c. Your question will be answered by one of the many volunteer "Math Doctors" from Swarthmore University. (Due to time restraints, you will have to periodically check back to this site for an answer to your question.)

d. You can also search through questions (and appropriate answers) that have been sent in by other students.

Reading, sports, geography, too,
Name an area that interests you!

Internet Idea #15

There is a web site for almost every area of interest. List some of your areas of interest. Use a search engine to find the many web sites that apply to your topic. Narrow down the search to highlight the web site that is most appropriate for your area of interest. You will find descriptions, data, and delight as you find that the Internet is full of words, pictures, and numbers for you to treat mathematically. Use percents, statistics, mapping skills, and more to observe and understand the information.

Internet Idea #16

Plan a trip! Decide where you would like to travel, when you want to leave and return, and your point of departure. Search the following web sites:

 http://www.travelocity.com

 http://www.expedia.com

 http://www.deltaairlines.com

and other airline web sites. Find the best value for the time of your travel and for your money. Decide if the least expensive fare is always the better fare in terms of time and convenience. Visit a travel agent to see how your trip could be a planned by a *professional*. Ponder whether the use of the Internet makes for a better trip.

Internet Idea #17

Look at the following web site:

http://www.census.gov/stat_abstract/img/wage.gif

Ponder the following questions:
1. Why do you think that a survey was conducted for this information?
2. Observe the data and the graphs and make a conclusion about the average earnings of year-round full-time workers in 1992 with regard to gender and/or education.
3. Do you think that every male with a Bachelor's degree earned $52,920 in 1992? What type of statistic, mean, median, or mode, do $52,920 and all of the other numbers in the graphs represent? Why do you think that this is so?
4. Find the range of the data for the males and then for the females.
5. Compare the earnings of people who attended or graduated college to those who did not. Does the comparison support your original conclusion about the graphs?
6. How do you think that this data was tabulated? Do you think that it is a good representation of people's earnings? Why or why not?

Internet Idea #18

Go to

http://www.askjeeves.com

and find the dimensions of the fields of your favorite baseball teams. Compare the areas and the distance a baseball needs to travel to reach a home run. In which parks is the home run distance the least and/or the most? Compare the area and perimeter of a baseball field to those of a football field, a volleyball field, and a soccer field. Find and compare the circumferences of a baseball, softball, volleyball, and a soccer ball.

Internet Idea #19

Find the price, with the sales tax, of one of your favorite books at a local bookstore. Go to

 http://www.amazon.com

or

 http://www.kozmo.com

and track down the web price for your book, including the shipping. Find the dollar difference and the percent difference. Determine the better buy in terms of money, time, convenience and/or enjoyment of an actual bookstore.

> Real-life problems
> Can be solved with no pain,
> Jump aboard and try them
> On the Internet train.

BRAIN TICKLERS
Set # 25

- Go to the listed web site.
- Read each question and search for the appropriate data needed for its solution.
- Plan a mathematical strategy to solve the problem.

1. **http://www.epicurious.com**

a. Go to *Recipe Search* and type in "Chocolate Chip Cookies." Click on **Go**.

b. Find the amount of flour that will be needed to make about two dozen Chocolate Chocolate Chip Cookies.

c. Find the number of cups of butter needed for four dozen Chocolate Chocolate Chip Cookies.

2.

http://www.usps.gov

Highlight *Postage Rates & Fees,* and click on **Calculate Domestic Postage**. Click on **A Letter**.

a. You are planning to send 125 one-ounce letters from your address at 250 Wireless Boulevard, Hauppauge, New York to your many friends at 1600 Pennsylvania Avenue, Washington, D.C.

* Search for the zip codes of each address.
* Find the difference in cost for mailing the letters separately via Priority mail compared to the cost for first-class mail.

b. Find the cost of mailing the letters overnight (Express).

3.

http://www.cyberdiet.com

* Click **Fast Food Quest**.
* Select **Pizza Hut** and **Domino's Pizza**.
* Select **Pizza** and **Pan/Deep Dish**.

a. Find the percent difference (compared to the Pizza Hut pizza) between the calories in a Domino's Pizza, Cheese Only, 6 inches, Ultimate Deep Dish and a Pizza Hut, Cheese Personal Pan.

b. Find the difference in the calories from fat percentages of the two pizzas.

4.

http://www.dartmouth.edu/~chance/

* Click on **Teaching Aids**.
* Click on **Data**.

a. Find the median income of a Boston Red Sox player in the 1994 season.

b. Find the difference in the range of salaries of the Baltimore Orioles players and the range of salaries of the Boston Red Sox players.

5.

http://www.census.gov

- Go to *People*, and click on **Estimates**.
- Click on **State**.
- Under *Maps*, click on **1998 to 1999**.
- Click on the U.S. map.

a. Find the mean percent population change of the six New England states from 1998 to 1999.

b. Find the mode percent change of all 50 states. In which states does the mode occur?

(Answers follow.)

BRAIN TICKLERS—THE ANSWERS

Set # 25, page 284

1. a. To bake about 36, or 3 dozen, cookies, $1\frac{1}{4}$ cups of flour are

needed. Let c stand for the number of cups needed for

2 dozen cookies. Use the proportion $\dfrac{1\frac{1}{4} \text{ cups}}{3 \text{ dozen}} = \dfrac{c \text{ cups}}{2 \text{ dozen}}$.

Cross multiply and divide to get $2.5 = 3c$; $2.5 \div 3 = c$;

$c =$ about 0.8 or $\frac{4}{5}$ of a cup of flour.

b. To bake 3 dozen cookies, $\frac{1}{2}$ cup of butter is needed. Use

the proportion $\dfrac{\frac{1}{2} \text{ cup}}{3 \text{ dozen}} = \dfrac{c \text{ cups}}{4 \text{ dozen}}$. Cross multiply and

divide to get $2 = 3c$; $2 \div 3 = c$, $= \frac{2}{3}$ of a cup of butter.

2. a. By doing a web search, you will find that the zip code for Hauppauge, New York is 11788 and the zip code for the given address in Washington, D.C. is 20500. Using the given web site, you will find that for first-class delivery it costs $0.34 to send one one-ounce letter, and that for Priority mail it costs $3.50. The difference is $3.50 – $0.34 = $3.16 per letter. For 125 letters, there will be a 125 × $3.16 = $395 difference.

    b. The cost of Express delivery for a one-ounce letter is $12.25. It would cost 125 × $12.25 = $1531.25 to mail the letters overnight.

3. a. There are 598 calories in the Domino's pizza and 813 calories in the Pizza Hut pizza. The difference is 813 – 598 = 215 calories. The Domino's pizza has $\frac{215}{813} \approx 26\%$ fewer calories.

    b. Using the fat gram data in the chart, you will find that the Domino's pizza has 27.6 fat grams and the Pizza Hut pizza has 27 fat grams. Either by using the data in the chart or by remembering that there are 9 calories per one gram of fat, you will find that the Domino's pizza has (27.6)(9) / 598 ≈ 42% fat calories and the Pizza Hut pizza has (27)(9) / 813 ≈ 30% fat calories. There is a difference of about 42% – 30% = 12% in percentages of fat. Remember that even if one pizza weighs more than the other you can still use the percentages of fat to compare them.

4. a. There were 25 players on the 1994 Red Sox team. You will see that the salaries are listed from highest to lowest. To find the median of 25 data values, look at the middle value, the thirteenth value (12 values above and 12 values below). Count from the highest salary ($5,155,250) to the thirteenth salary listed, $650,000. The median salary was $650,000.

    b. To find the range, subtract the lowest salary from the highest salary. For the Baltimore Orioles, the range was $5,406,603 – $109,000 = $5,297,603. For the Red Sox, the range was $5,155,250 – $109,000 = $5,046,250. The difference in the ranges was $5,297,603 – $5,046,250 = $251,353.

5. a. The six New England states are Connecticut, Maine, Massachusetts, New Hampshire, Rhode Island, and Vermont. Their percent population changes were, in order, 0.3%,

0.4%, 0.5%, 1.3%, 0.3%, and 0.5%. Their mean is $(0.3 + 0.4 + 0.5 + 1.3 + 0.3 + 0.5) \div 6$, which simplifies to $3.3 \div 6$, or a mean of 0.55%.

b. Looking through the map for common percents of change, you will find that 0.6% appears the most often, at six times. Thus, the mode percent change is 0.6%. The states of Kansas, Oklahoma, Missouri, Mississippi, Indiana, and New Jersey all had a percent change of 0.6%.

INDEX

NOTES

NOTES

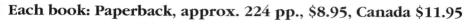